HOW TO MAKE AN
APPLE PIE FROM SCRATCH

Harry Cliff is a particle physicist based at the University of Cambridge and was a curator at the Science Museum in London for seven years. He regularly gives public lectures and makes TV and radio appearances. His 2015 TED talk 'Have We Reached the End of Physics?' has been viewed more than 2.5 million times. He lives in London.

HOW TO MAKE AN

APPLE PIE

FROM SCRATCH

In Search of the Recipe for Our Universe

HARRY CLIFF

PICADOR

First published 2021 by Picador

This edition published 2022 by Picador
an imprint of Pan Macmillan
The Smithson, 6 Briset Street, London EC1M 5NR
EU representative: Macmillan Publishers Ireland Ltd, 1st Floor,
The Liffey Trust Centre, 117–126 Sheriff Street Upper, Dublin 1, D01 YC43
Associated companies throughout the world
www.panmacmillan.com

ISBN 978-1-5290-2621-4

A CIP catalogue record for this book is available from the British Library.

Printed and bound by CPI Group (UK) Ltd, Croydon, CR0 4YY

Visit **www.picador.com** to read more about all our books
and to buy them. You will also find features, author interviews and
news of any author events, and you can sign up for e-newsletters
so that you're always first to hear about our new releases.

For Vicky and Robert.
Thank you.

If you wish to make an apple pie from scratch, you must first invent the universe.

— Carl Sagan

Contents

HOW TO MAKE AN

APPLE PIE

FROM SCRATCH

Prologue

On a frosty morning in March 2010 I pulled up outside a fenced compound on the outskirts of the French commune of Ferney-Voltaire. A sign bolted to the steel security gates announced

CERN SITE 8
ACCÈS RÉSERVÉ AUX PERSONNES AUTORISÉES

Leaning awkwardly across to reach through the passenger window of my right-hand-drive car, I swiped my security badge against the reader. The gates remained closed. Hmmm . . . had my access request not gone through? Noticing a queue of cars beginning to form behind me, I gave the reader a series of increasingly anxious swipes. Nothing. I was just about to get out to attempt to negotiate with the security guard in my halting high-school French when, to my relief, the gates began to creak open.

I parked behind the main experimental hall, facing the chain-link fence that marks the boundary of Geneva Airport's runway. Outside, my breath misted in the cold air, which carried a now-familiar sickly sweet smell from a perfume factory in the nearby Swiss town of Meyrin. Pushing my hands into my coat pockets I made for the prosaically named Building 3894, a single-storey portacabin used for the early morning run meetings.

Inside, most of the participants were already crowded around the long table waiting for the meeting to start. Some chatted with their neighbours in English, French, German, Italian; others sipped coffees or sat hunched over laptops. I took my seat a row back from the table itself, hoping that I wouldn't be called upon.

A hundred metres beneath our feet in a concrete tunnel so long it could encircle a city, the largest and most powerful machine ever built was being coaxed into life: the Large Hadron Collider (LHC). In just a few days, this ring-shaped particle accelerator would slam subatomic particles into one another with such incredible violence that it would briefly recreate conditions that existed during the first instant after the big bang.

These tiny cataclysms would be recorded by four giant particle detectors, housed in cathedral-sized underground caverns, spaced several kilometres apart around the LHC ring. One of these detectors was directly below us – the Large Hadron Collider beauty (LHCb) experiment – 6,000 tonnes of steel, iron, aluminium, silicon, and fibre-optic cables, poised like a sprinter in the blocks, waiting for its moment to arrive.

It had been a long wait. Some of my colleagues had spent their entire careers building towards this moment. Twenty years of planning, funding bids, scrupulous design, testing, and engineering had resulted in one of the most advanced particle detectors ever built. In the next few days all that work would finally be put to the test, as engineers on the LHC prepared to collide particles inside the detector for the first time.

I was twenty-four years old, a second-year PhD student, having arrived in Geneva for the first of two three-month stints a few weeks earlier. My new home was CERN, the European Organization for Nuclear Research, the largest and most advanced particle physics lab in the world. Over the past few weeks I had slowly learned to find my way around the labyrinth of office buildings, workshops, and laboratories that make up the sprawling CERN site, battled through February snowstorms, and discovered that flushing your toilet after ten p.m. in Switzerland will get you a stern telling-off from your neighbours. I was also getting to grips with my new duties on LHCb, including responsibility for one of its numerous subsystems, each of which would have to function flawlessly. If one failed, then the long-awaited data could end up being unusable.

I had first come face-to-face with LHCb a year and a half earlier. My supervisor, Uli, a German postdoctoral researcher who was based at CERN full-time, had guided me through the complex set of procedures required to access the detector. Donning a badge that would monitor my radiation exposure during my trip below ground, I first had to persuade a rather temperamental iris scanner to let me through a set of bright green airlock-style security doors. Then a small metal lift shuddered its way 105 metres beneath the earth, down into what is rather ominously known as 'the pit'.

The doors opened on a strange subterranean world of whirring machinery, metal gantries painted in primary colours, and concrete tunnels threaded with thousands of metres of cables and ducts. Another set of security doors, this time bright yellow and emblazoned with radiation warning signs, and then a narrow passage snaking its way through a 12-metre-thick shield wall before abruptly opening into a soaring concrete cavern.

The first thing that strikes you is its sheer size. LHCb is big: 10 metres high and 21 metres long, spanning the entire width of the cavern. At first glance it can be hard to figure out what you're looking at; the view is dominated by staircases, steel platforms, and scaffolding, painted in green and yellow, that support and allow access to the sensitive elements of the detector, which are mostly hidden from view. Criss-crossing the walls of the cavern are reams of cables taking power to the detector and carrying away the torrent of data produced by millions of tiny, precision-engineered sensors. LHCb is capable of measuring the paths of thousands of individual subatomic particles as they tear out from the collisions at a whisker below the speed of light with a precision of a few thousandths of a millimetre. And it can do this a million times every second.

But perhaps the most remarkable thing about LHCb is the way it was built. Like all four of the large LHC experiments, it is a modern-day Tower of Babel, with each component designed and assembled by an international collaboration of physicists and engineers based at dozens of universities spread across the globe, from Rio de Janeiro to Novosibirsk. Brought together in this giant hole in the ground just outside Geneva, they form a single, mind-meltingly complex instrument. The fact that any of this works at all still seems kind of miraculous to me.

My colleagues in Cambridge had spent the last decade designing, building, and testing the electronics that would read out data from the subdetector whose job is to tell different types of particle from one another. My small part in all this was to make sure that the software used to control and monitor the electronics worked without crashing or otherwise causing problems when the moment came. I was a small cog in a huge machine, but still, I was acutely aware that two decades of effort by hundreds of physicists from seventy countries and an investment of €65 million from more than a dozen national funding agencies depended on me doing my small job properly. I did not want to be the person who fucked up at the last minute.

The chatter in the room ceased abruptly as the run chief called the meeting to order. I glanced around the room at my colleagues, many of whom looked as though they hadn't had much sleep in the past few days, aware that this was the beginning of the most important phase of my career so far. The first item was a report detailing overnight work on the LHC, which people at CERN colloquially refer to as 'the Machine'. It was this machine that we were all now waiting for.

More than three decades in the making, the LHC is a scientific project on an unprecedented scale. Almost everything about it is extreme. It's the largest scientific instrument ever built, by some measures the largest machine ever built: 27 kilometres in circumference, so large that it crosses the border between France and

Switzerland four times (there are actually flags marking the border painted on the tunnel walls). The beam pipes that carry the particles are emptier than interstellar space, while the thousands of superconducting magnets that steer the particles around the ring operate at the staggeringly low temperature of –271.3 degrees Celsius, less than 2 degrees above absolute zero. To achieve this requires the world's biggest cryogenic facility, which uses 10,000 tonnes of liquid nitrogen and as much electricity as a large town to produce over 120 tonnes of superfluid liquid helium, which is then pumped intravenously through the LHC's magnets. Within a few days, this giant machine would start accelerating subatomic particles called protons to 99.999996 per cent of the speed of light, before firing them headlong into one another at four points around the ring, including inside LHCb, creating forms of matter not seen in large quantities since a trillionth of a second after the universe began.

All of this, the years of design work and funding negotiations, the mobilization of a global community of thousands of physicists, the civil engineering (which included digging through an underground river that had been frozen using liquid nitrogen), not to mention manufacturing, testing, and installing millions of individual components, from 35-tonne magnets to the tiniest silicon sensors, was to serve one cause: curiosity. Despite what some tabloids might try to tell you – for instance the *Daily Express* never seems to tire of suggesting that CERN is using the LHC for nefarious purposes, including opening a portal to another 'sinister' dimension (perhaps that gateway to 'the Upside Down' in *Stranger Things* was really CERN's fault) or, my all-time favourite, 'to summon God' – the LHC exists only to answer fundamental questions about the most basic building blocks of our world and how our universe came into being.

And there are some really big questions that we need answers to. Our current theory of what the world is made from down at the fundamental level is known as the 'standard model' of particle physics – a deceptively boring name for one of humankind's greatest intellectual achievements. Developed over decades through the combined efforts of thousands of theorists and experimentalists, the standard model says that everything we see around us – galaxies, stars, planets, and people – is made of just a few different types of particles, which are bound together inside atoms and molecules by a small number of fundamental forces. It's a theory that explains everything from why the Sun shines to what light is and why stuff has mass. What's more, it's passed every experimental test we've been able to throw at it for almost half a century. It is, without a doubt, the most successful scientific theory ever written down.

All that said, we know that the standard model is wrong, or at the very least seriously incomplete. When it comes to the deepest mysteries facing modern physics, the standard model simply shrugs or offers up a bunch of contradictions instead of answers. Take this for starters. After decades of painstakingly peering into the heavens, astronomers and cosmologists are pretty well convinced that 95 per cent of the universe is made of two invisible substances known as 'dark energy' and 'dark matter'. Whatever they are – and to be clear we haven't got much of a clue about either of them – they're definitely not made from any of the particles in the standard model. And as if missing 95 per cent of everything wasn't bad enough, the standard model also makes the rather startling assertion that all the matter in existence should have been wiped out in a cataclysmic annihilation with antimatter in the first microsecond of the big bang, leaving a universe with no stars, no planets, and no us.

So it's pretty obvious that we are missing something big, most likely in the form of some as-yet-undiscovered fundamental particles that could help explain why the universe is the way it is.

Enter the Large Hadron Collider. As we sat gathered around that meeting table in March 2010, there was huge optimism that we'd soon spy something altogether new or unexpected come flying out from the collisions produced by the LHC. If that happened, then it would be the start of a process that could help unravel some of the biggest mysteries in science.

When I signed up for my PhD in early 2008, I knew that I'd be starting out in particle physics just as the LHC switched on for the first time. I was thrilled by the idea of being among the very first students to see data from a machine that had been in development since the late 1970s and had cost more than €12 billion. On 10 September 2008, just a few days before I arrived at my new lab in Cambridge, the LHC was launched in a blaze of publicity. Under the glare of the world's media, protons were sent around the 27-kilometre ring for the first time. Champagne bottles popped as physicists and engineers celebrated one of the greatest scientific feats in history, and particle physics was briefly headline news.

A few days later, the LHC was back in the news for a different reason. At around midday on 19 September, during final tests of the collider's electromagnets, something catastrophic happened. Engineers in the LHC Control Centre, CERN's equivalent of NASA's Mission Control, watched in disbelief as screen after screen all around the huge room turned lurid red. An engineer I spoke to later told me that at first there were so many alarms going off that they thought there must be something wrong with the software used to monitor the accelerator. Hours later, when they finally made it down into the tunnel, he and his colleagues were confronted with a scene of devastation.

A single loose connection had caused an electrical arc that flash boiled the bath of liquid helium used to cool the magnets, creating a shockwave that sent a cascade of destruction along a 750-metre stretch of the accelerator. Fifteen-metre-long electromagnets weighing up to 35 tonnes had been torn from their moorings and shunted across the tunnel. The faulty connection itself had been vaporized, blasting black soot hundreds of metres down the ultra-clean beam pipes in both directions.

Repairs would take more than a year. Despite an initial loss of confidence, the engineering staff at CERN soon dusted themselves off and got back to work. On 20 November 2009, fourteen months and €25 million later, they tentatively sent protons back around the LHC for the first time since what is now euphemistically referred to as 'the incident'. However, that had only been a dry run, with the accelerator coasting at a small fraction of its maximum energy.

Now, in March 2010, we were finally approaching the moment when the machine would be pushed into uncharted territory, reaching collision energies that would allow us to begin to search for dark matter, the Higgs boson, microscopic black holes, and perhaps other exotic objects that no one had yet imagined. I suspect everyone sitting around the table that morning felt the weight of what we were about to do.

The run chief gave his report, pausing occasionally when he was drowned out by the roar of a passenger jet taking off from the nearby runway. Aside from a brief power failure, overnight work on the LHC had gone smoothly and we were on track to see collisions within a few days. He then moved around the table, as physicists from the Netherlands, Spain, Russia, Germany, and Italy gave updates on their subsystems in perfect English. There was a brief Eurovision moment when a French physicist launched into his report in his native tongue. Despite a bit of eye-rolling from around the table, the physicist ploughed stubbornly on, not unjustified really given that French is one of the two official languages of CERN, and what's more that we were in France. That said, almost all meetings at CERN are conducted in English and my French wasn't quite up to the task of following what I assume was a technical discussion of some aspect of the experiment.

I could feel my heart beating a little faster as my turn approached. We had had one minor problem with the software that controlled the electronics a few days earlier, triggering a panicked rush to the control room at the crack of dawn. Eventually the problem had been fixed using the classic solution – turn it off and on again – and all had been running smoothly since. But at the back of my mind the fact that I hadn't tracked down the root cause of the error was nagging at me.

'Nothing to report over the last twenty-four hours,' I said, hoping that there would be no follow-up questions. To my relief, the run chief's attention turned to the next subsystem and after a few more short reports the picture was clear: LHCb was ready.

Outside in the car park I watched clouds of steam billowing from the cooling towers, the only visible evidence of the huge machine that waited below. I wondered for a moment how many of the residents of that stretch of countryside between Geneva Airport and the Jura Mountains were aware of what was going on beneath their feet.

A little more than a week later, on 30 March 2010, engineers at the LHC pulled off the spectacular feat of firing two beams of protons at each other and getting them to collide head-on, which is more or less equivalent to launching two knitting needles at each other from opposite sides of the Atlantic and getting them to hit halfway. As the first protons collided, energy gave birth to matter, and screens around CERN lit up with images of that first microscopic moment of creation. The physicists crammed into the small LHCb control room erupted in cheers and applause. The work of two decades had finally paid off.

That day marked the beginning of a bold new phase in humankind's most ambitious intellectual journey: the centuries-long quest to uncover nature's most basic ingredients and to figure out where they came from, what you might call the search for the recipe for our universe. This book is the story of that quest. It's the story of how thousands of people working over hundreds of years gradually discovered the fundamental ingredients of matter and traced their origins out into the cosmos, through the hearts of dying stars and back to the first furious moments of the big bang. It's a story that takes in chemistry, atomic, nuclear, and particle physics, astrophysics, cosmology, and more besides, and it's a story that I will tell through my personal mission to find the ultimate recipe for apple pie. Why an apple pie, you ask? Well . . .

In the landmark television series *Cosmos*, the American astrophysicist Carl Sagan took audiences on an epic journey through the universe, flying to distant galaxies, seeking out the origins of life, and witnessing the births and deaths of stars. And as *Cosmos* was made in 1980, this voyage through space and time was accompanied by a lot of synth.

Sagan, who sometimes got made fun of for his rather portentous presenting style, engaged in a bit of self-satire in episode 9, which begins with what at first glance appears to be a small green planet floating in the vacuum of space. As we fly closer, we realize it's not a planet after all, it's an apple, which suddenly gets sliced in two as we cut to a kitchen scene where a rather ominous-looking rolling pin dramatically flattens a ball of dough, all to a swelling score that could be straight out of *Blade Runner*.

The sequence ends in the grand oak-panelled dining hall of Cambridge's Trinity College, where Sagan, looking rather dapper in one of his signature red turtleneck sweaters, is seated at the head of a long table. A waiter presents him with a freshly baked apple pie, and Sagan turns to camera with a twinkle in his eye and says, 'If you wish to make an apple pie from scratch, you must first invent the universe.'

Now that's a cooking show I'd like to watch. 'Today on *The Great British Bake Off* we're going to be making salted caramel parfait, but first Mary Berry is going to show you how to synthesize carbon using a dying star.' Anyway, Sagan's point was that an apple pie is far more than just apples and pastry. Zoom in far enough and you'll discover trillions and trillions of atoms, which were blasted into space by supernovae or forged in the searing heat of the big bang. So if you really want to understand how to make an apple pie, you need to figure out how to make the entire universe.

Understanding the ultimate origin of everything is usually put in more grandiloquent terms – Stephen Hawking famously described it as knowing 'the mind of God' – but I rather like Sagan's more down-to-earth take. If we start with an apple pie and break it down into ever-more-basic ingredients, while at the same time trying to figure how they were made, will we eventually reach an end point? We may never know the mind of God, but might we be able to figure out how to make an apple pie from scratch?

Getting an answer to that question will take us on a journey across the globe, plunging a kilometre beneath an Italian mountain range to peer into the heart of our Sun, and climbing to the top of a high New Mexican peak where astronomers decode signals hidden in starlight. We'll listen to ripples in the fabric of space and time amid the humid pine forests of southern Louisiana and go behind the scenes at the New York lab where a giant particle collider recreates temperatures not seen since the big bang. Along the way, we'll cross paths with chemists, astronomers, physicists, and cosmologists, past and present, on a quest to uncover the fundamental ingredients of matter and reveal their histories. And we'll face up to the mysteries that remain unsolved and ask whether there are questions we may *never* be able to answer.

We'll cross continents and centuries in pursuit of the recipe for our universe, but like all epic sagas, this journey begins at home.

CHAPTER 1

Elementary Cooking

One summer afternoon, I arrived at my parents' house in suburban south-east London armed with some glassware that I'd ordered online and a pack of six Mr Kipling Bramley apple pies. I was there to do what is probably the silliest experiment I've ever attempted.

As a child, my dad was a keen amateur chemist and used to spend happy afternoons in the mid-1960s creating smells and explosions in the shed at the bottom of his parents' garden. Those were the days when anyone (including teenagers in possession of an advanced knowledge of chemistry and a healthy disregard for their own safety) could buy a terrifying array of noxious substances from their local chemical supplier. This, it turned out, included all the ingredients of gunpowder. He still recalls with some relish how one of his more dramatic experiments was brought to an abrupt end when his own father, a former artilleryman not unaccustomed to the sound of gunfire, stormed to the bottom of the garden shouting, 'That's enough, that one rattled the windows!' Simpler times. My dad still has some of his old chemistry equipment, including a Bunsen burner that I wanted to get my hands on, and I'd decided that my small London flat was probably not the ideal location for the experiment I had in mind.

The thought behind the experiment was this: if you were

presented with an apple pie and had no knowledge of pies, apples, or their composition, what might you do to try to figure out what it was made from? On the workbench in the garage I scraped a small sample of the pie into a test tube, taking care to get a good mix of the crumbly pastry and the soft apple filling, and then sealed it with a cork that had a small hole drilled through the middle. After connecting the tube to a second flask floating in a tub of cold water via a long L-shaped glass pipe, we fired up the Bunsen burner, popped it under the test tube, and stood back.

The pie began to bubble and caramelize, and soon the expanding gas within the test tube threatened to force our sample up into the connecting pipe. Reducing the heat slightly we watched the pie slowly start to blacken, and to my delight tendrils of mist started to flow along the pipe and pour into the waiting flask, which before long was overflowing with a ghostly white vapour. Now this was a real chemistry experiment!

Wondering what this white mist might be, I gave it a whiff, a tried and tested method of chemical analysis from before the days of health and safety. Humphry Davy, a pioneering chemist of the Romantic age, famously investigated the medical effects of various gases by inhaling them, which in 1799 led him to discover the pleasurable effects of nitrous oxide, what we now know as laughing gas, which he would inhale in large quantities while locked in a dark room with his poet friends, or sometimes young women of his acquaintance. Mind you, it wasn't a risk-free strategy. He came close to killing himself during an experiment with carbon monoxide, and on being dragged into the open air remarked faintly, 'I do not think I shall die.'

Alas, my apple pie vapour didn't produce any psychoactive effects, just an extremely unpleasant burned smell that seemed to hang around for hours afterwards. Peering through the mist to the bottom of the flask I found that some parts of the vapour had condensed on contact with the cool water bath, forming a yellowish liquid covered by a dark brown oily film.

After about ten minutes of intense heating, no more vapour seemed to be coming off the charred remains of the apple pie and so we concluded that our experiment was complete. In my keenness to inspect the contents of the test tube, I briefly forgot that when you heat glass with a Bunsen flame for ten minutes it gets really quite hot and badly burned my index finger. There's a good reason why the most dangerous bit of equipment I am generally allowed near is a desktop computer.

After a much longer wait, I gingerly returned to the test tube and tipped its contents onto the bench. The apple pie had been reduced to a jet-black, rocklike substance whose surface was slightly shiny in places. So what can we conclude about the composition of apple pie from this admittedly rather silly experiment? Well, we've ended up with three different substances: a black solid, a yellow liquid, and a white gas, which by now had infused my skin, hair, and clothes with a nauseating burned smell. I admit that the precise chemical composition of these three apple pie components was not entirely clear to me at the time, though I was pretty sure the black stuff was charcoal and that the yellowish liquid was probably mostly water. To get further towards a list of fundamental apple pie ingredients we are going to need to do some more advanced chemical analysis.

THE ELEMENTS

I shouldn't admit this as a physicist, but chemistry was my favourite subject at school. Physics labs were sterile, joyless places where we were expected to find excitement wiring up a circuit or glumly timing the swing of a pendulum. But the chemistry lab was a place of magic, where you could play with flame and acid, set fire to magnesium ribbon that burned so bright it dazzled, or bubble coloured potions through delicate glassware. The safety glasses, the bottles of sodium hydroxide with threatening orange warning labels, and white lab coats stained with the unidentified, perhaps toxic, remains of experiments past, all helped to lend the chemistry lab a frisson of danger. And marshalling all this was our enigmatic teacher, Mr Turner, who arrived at school in a sports car and was rumoured to have made his fortune by inventing the spray-on condom.

In fact, it was a fascination with chemistry that set me on a path towards eventually becoming a particle physicist. Chemistry, like particle physics, concerns itself with matter, the stuff of the world, and how different basic ingredients react, break apart, or change their properties according to certain rules. The reason I didn't stick with chemistry in the end is because I wanted to know where those rules came from. Had I been born in the eighteenth or nineteenth century, I would most likely have stuck with it. Back then, if you wanted to understand the fundamental building blocks of matter, then chemistry, not physics, was the subject for you.

The person who probably did more than anyone else to invent modern chemistry was Antoine-Laurent Lavoisier, a brash, ambitious, and fabulously rich young Frenchman who lived and worked in the second half of the eighteenth century. Born in Paris in 1743 into a wealthy family steeped in the legal profession, he used a large inheritance from his father to equip his personal lab at the Paris Arsenal with the most sophisticated chemical apparatus money could buy. Aided by his wife and fellow chemist, Marie-Anne Pierrette Paulze, he brought about a self-declared 'revolution' in chemistry by systematically dismantling the old ideas that had been inherited from ancient Greece and inventing the modern concept of the chemical element.

The idea that everything in the material world is made up of a number of basic substances, or elements, has been around for thousands of years. Different element theories can be found in ancient civilizations, including Egypt, India, China, and Tibet. The ancient Greeks argued that the material world was made of four elements: earth, water, air, and fire. However, there is a big difference between what the ancient Greeks thought of as an element and the definition of a chemical element that we learn about at school.

In modern chemistry, an element is a substance like carbon, iron, or gold that can't be broken down or converted into anything else. On the other hand, the ancient Greeks thought that earth, water, air, and fire *could* be transformed into one another. On top of the four elements they added the concept of four 'qualities': hotness, coldness, dryness, and moistness. Earth was cold and dry, water was cold and moist, air was hot and moist, and fire was hot and dry. This meant that it was possible to convert one element into another by adding or removing qualities; adding hotness to water (cold and moist) would produce air (hot and moist), for example. This theory of matter raised the prospect of transforming, or 'transmuting', one substance into another – most famously common metals into gold – through the practice of alchemy.

It was the concept of transmutation that Lavoisier attacked first. As with many of his greatest breakthroughs, his approach was based on a simple assumption, namely that mass is always conserved in a chemical reaction. In other words, if you weigh all the ingredients at the start of an experiment, and then all the products at the end, taking care to make sure no sneaky wisps of gas escape, then their masses should be the same. Chemists had been making this assumption for some time, but it was Lavoisier, aided by a set of extremely precise (and expensive) weighing scales, who popularized the idea when he published the results of his own painstaking experiments in 1773.* In Mr Turner's chemistry lessons, the law of the conservation of mass was taught to me as Lavoisier's principle.

One piece of evidence in transmutation's favour was the fact

---

* In fact, the Russian polymath Mikhail Lomonosov had discovered the law of the conservation of mass in his own experiments many years earlier, but Lavoisier's tremendous influence over the development of modern chemistry means that poor old Lomonosov is mostly forgotten.

that when water was slowly distilled in a glass container, a solid residue was left behind, which seemed to confirm that water could be converted into earth. Lavoisier had his doubts. Weighing the empty glass container before and after the experiment, he found that it had lost some mass, which was almost exactly equal to the mass of the so-called earth. In other words, the idea was nonsense. The solid residue was just made up of bits of the glass container.

By demolishing the idea of the transmutation of water into earth, Lavoisier fired the first shot in a campaign that would totally upend how people thought about the chemical world. Declaring with characteristic swagger his intention to bring about 'a revolution of physics and chemistry', he then set about tearing down the elements themselves. His next move was to take on the most mysterious and powerful of them all: fire.

In the mid-eighteenth century, flammable materials like charcoal were believed to contain a substance known as 'phlogiston' that was given off when they were set on fire. A fuel like charcoal contained lots of phlogiston, which was released during burning, with the burning eventually stopping either when all the phlogiston in the charcoal had run out or when the surrounding air had become so full of phlogiston that it couldn't absorb any more.

One problem with this phlogiston business came with the discovery that metals actually get heavier when they are burned, whereas you'd expect them to get lighter if phlogiston was being released. This was explained away by the Dijon-based lawyer and chemist Louis-Bernard Guyton de Morveau as being due to the fact that phlogiston was incredibly light and when stored in metals somehow 'buoyed' them up, a bit like a hot-air balloon. When the metal was burned, the buoyancy provided by the phlogiston was lost and so the metal appeared to get heavier.

Lavoisier was less than impressed by Guyton's idea and argued the complete opposite – instead of burning releasing phlogiston, burning involved air being absorbed. This explained why metals got heavier when burned: they weren't releasing floaty phlogiston, they were combining with air.

It's worth taking a moment to appreciate how brilliant an insight this is. If you're briefly able to forget everything you were taught at school about combustion, then thinking that phlogiston is released by fire actually makes a lot of sense. Fire definitely seems to be a process that releases stuff – light, heat, and smoke at the very least. The idea that burning combines air with the fuel, effectively sucking something *out* of the air, is really quite counterintuitive. Lavoisier's ability to follow the experimental evidence and reject what might seem like common sense is what allowed him to leap to such a radically different conclusion.

The question was, what exactly was it in air that was consumed in burning? Unknown to Lavoisier at the time, significant advances in the understanding of air had recently been made across the Channel in Britain. In 1756 the Scottish natural philosopher* Joseph Black had discovered a peculiar new type of air that was released when certain salts were heated. Most surprisingly, he found that it was impossible to set things on fire when they were surrounded by this 'fixed air' – what we now know as carbon dioxide. A decade later, Henry Cavendish found that when sulphuric acid was poured over iron it gave off another, lighter air that would catch fire with a characteristic *pop*. But the most prolific discoverer of new airs was the English natural philosopher Joseph Priestley.

Priestley was inspired to begin his own investigations of air when he learned of Cavendish's discovery of 'inflammable air' in 1767. At the time he was working as a Presbyterian minister in Leeds and living next door to a brewery, a bit of a contrast to Lavoisier's lavishly equipped laboratory in central Paris. However, being next door to a brewery did have its benefits, aside from an ample supply

* People who studied the natural world were known as 'natural philosophers' until well into the nineteenth century, when the word 'scientist' first started to be used.

of beer. The fermentation process released large quantities of fixed air, which, among other things, Priestley used to develop a technique for making fizzy drinks, laying the foundations for the future soft-drink industry.*

A few years later, in 1774, Priestley made the discovery that would secure his place in the history books. He found that when he focused sunlight onto a sample of highly toxic 'red calx' (a mineral containing mercury) using a large burning lens, it gave off a new type of air that Priestley discovered would make a flame burn incredibly brightly and could keep a mouse in a sealed jar alive four times as long as normal. Priestley even tried the new air himself, writing,

> The feeling of it to my lungs was not sensibly different from that of common air; but I fancied that my breast felt peculiarly light and easy for some time afterwards. Who can tell but that, in time, this pure air may become a fashionable article in luxury. Hitherto only two mice and myself have had the privilege of breathing it.

Priestley believed the miraculous properties of what he called 'dephlogisticated air' were a result of it containing far less phlogiston than ordinary air. This allowed it to soak up the phlogiston released by a burning candle or a breathing mouse more effectively and thus keep them going for longer.

In October of that year, Priestley travelled to Paris, where he met many of the city's brightest minds, including Antoine Lavoisier. Unfortunately, we know very little about their meeting, but it's fun to imagine what these two chemical giants might have made of each other: the wealthy, self-confident and urbane Parisian and the

---

* Priestley never made any money out of his invention, but his technique was later taken up by J. J. Schweppe to make carbonated mineral water, founding the Schweppes Company in Geneva in 1783.

working-class radical with a strong Yorkshire accent. What we do know is that Priestley told Lavoisier about his new discovery, which proved to be the vital clue that he needed to complete his theory of fire. However, Lavoisier came to a radically different conclusion. Instead of dephlogisticated air, he realized that Priestley had in fact discovered the gas that combined with fuel during burning. He named it 'oxygen'.

According to Lavoisier, fire wasn't an element, and phlogiston didn't exist. When a candle burned, the fuel combined with oxygen to release carbon dioxide. Lavoisier showed that a similar process took place when animals breathed: carbon in their food combined with oxygen to release carbon dioxide and heat. He even demonstrated this idea using a guinea pig placed in an empty bucket surrounded by a container full of ice. The heat from the rodent's body melted the ice, and by measuring the amount of water that ran out of the bottom of the container Lavoisier was able to figure out how much heat it was giving off, proving that animals effectively burned their food to create heat. Don't worry, the guinea pig escaped freezing to death – though it definitely would have got a bit chilly – and has the possibly dubious honour of being the original source of the term 'to be a guinea pig'.

Lavoisier wasn't done with his revolution yet. People had noticed that when Cavendish's inflammable air was burned with oxygen, water seemed to be left behind. Lavoisier became convinced that this meant that water, once thought to be the most basic of all the elements, wasn't an element either. Instead it was made from this inflammable air, which he renamed 'hydrogen', and Priestley's oxygen.

Most of the scientific community, particularly in France's great imperial rival Britain, found it hard to swallow Lavoisier's radical new ideas. Priestley rejected Lavoisier's suggestion that water wasn't an element and clung to phlogiston theory for the rest of his life. Lavoisier needed firm experimental proof to swing people behind his new chemistry. He finally provided it in spectacular fashion by splitting water into oxygen and hydrogen in a public demonstration held at his laboratory in 1785.

By the late 1780s the old classical elements lay in ruins. Water could be broken apart into hydrogen and oxygen, air was a mixture of different gases, and fire was a process of combining oxygen with fuel. In 1789 Lavoisier published his greatest piece of propaganda for the new chemistry: a textbook called *Traité élémentaire de chimie* (*An Elementary Treatise on Chemistry*). In it, he gave his new definition of a 'chemical element', a substance that couldn't be broken down into anything else. What's more, he provided a list of thirty-three of these new chemical elements, many of which we still recognize today, including oxygen, hydrogen, and azote, or what we would now call nitrogen. The treatise became one of the most influential books in the history of science and within a few years all but his most die-hard critics had been won over. Lavoisier had lived up to his arrogant claim; he really had brought about a chemical revolution.

So what might Lavoisier have made of the three products of my apple pie experiment? First of all, I suspect he would have been rather unimpressed by my rough and ready approach to chemistry. My dad's garage isn't quite as well equipped as Lavoisier's lab and I didn't have any kit that would have allowed me to weigh the apple pie precisely before and after the experiment as Lavoisier surely would have done. Worse still, I'd carelessly allowed the white mist to escape, meaning that its composition would have to remain a mystery.

But what of the charred black hard stuff left behind in the test tube? If we take a look at Lavoisier's list of chemical elements one jumps out immediately: charcoal. Charcoal has been used as a fuel for centuries and was often made by burying piles of wood under a layer of turf and lighting a fire in the centre. The turf kept the air out, preventing the sides of the woodpile from catching on fire, while the intense heat from the central fire broke the wood down into charcoal and gases. This is more or less what we had done with the apple pie; the bung in the test tube had acted like the turf, stopping the oxygen in the air from getting in and preventing the superheated pie from catching fire. We had made charcoal. Or what in modern terms is a fairly pure form of the basic element of all organic matter: carbon.

As for the yellowish liquid, well, in principle I could have tried to break it down further, but unfortunately I'd only been able to generate a thimbleful of the foul-smelling liquid – far too little for the experiment to work – and I wasn't about to empty my local supermarket of apple pies and spend days bubbling them down. Anyway, it seems a safe bet that it was mostly water and thanks to Lavoisier we know that water is a compound of oxygen and hydrogen, giving us another two ingredients. Indeed, carbon, oxygen, and hydrogen between them are the dominant chemical elements in all organic matter, from apple pies to humans. However, they are certainly not the only chemical ingredients. A quick glance at the nutritional information on the back of the box tells me that they contain at least some iron, which was probably still mixed up in the charcoal. And although I couldn't isolate them in my dad's garage, there's also nitrogen, selenium, sodium, chlorine, potassium, calcium, phosphorus, fluorine, magnesium, sulphur, and probably many more – perhaps only in very tiny amounts, but they're there.

The deeper question though is this: What are these different chemical elements made of themselves? After all, if we really want to make an apple pie from scratch, hydrogen, oxygen, and carbon won't cut it. They are just the start of the story.

CHAPTER 2

The Smallest Slice

A t the start of *Cosmos* episode 9, just after uttering the immortal phrase that inspired this book, Carl Sagan gets up from his seat at the head of the grand table and picking up a knife poses us a question: 'Suppose I cut a piece out of this apple pie . . . and now suppose we cut this piece in half, or more or less, and then cut this piece in half, and keep going . . . How many cuts before we get down to an individual atom?'

Ten? A hundred? A million? Perhaps you can keep cutting up the apple pie forever into ever smaller and smaller pieces, until you have an infinity of infinitesimal slices. This neat little thought experiment captures the essence of the most powerful idea in science – that everything is made of atoms.

Atoms, according to the classic definition, are tiny, indestructible nuggets of matter that can't be changed or broken apart (the word 'atom' comes from the ancient Greek *atomos*, meaning 'uncuttable'). They come in different shapes and sizes and combine to create everything we see in the world around us, from apple pies to astronauts. It's a beguilingly simple idea and yet at the same time goes completely against our everyday experience. Our senses reveal a world of form and colour, texture and temperature, taste and smell: the smooth red skin of an apple or the bitter taste of coffee.

Atomic theory tells us that this world is an illusion. Deep down at the roots of things there is no such thing as the colour red or the taste of coffee. Deep down there are only atoms and empty space. Colour, taste, heat, texture are all tricks of the mind that emerge from uncounted multitudes of different atoms, bound together in a dazzling array of different forms.

When you think about atoms this way it's not surprising that the idea took millennia to take hold. Although versions of atomic theory appeared in ancient Greece, they never really gained much traction, particularly as the influential Aristotle dismissed the idea, preferring to trust his senses over abstract thinking: The theory of qualities makes far more sense; we're all familiar with hotness, coldness, dryness, moistness, but who of us has ever seen an atom?

It was only in the seventeenth century that atoms started to be taken seriously in scientific circles. Isaac Newton was an avowed atomist and believed that atoms made up not only the material world but even light itself, which he imagined as a shower of tiny particles or 'corpuscles'. Newton's mighty legacy to science, along with gravity, optics, and the laws of motion, included persuading many eighteenth-century natural philosophers to take an atomic view of the world. That said, there was precious little evidence for atoms' existence, and the concept was pretty useless for understanding chemistry. Lavoisier and Priestley could experiment and theorize without having to worry very much about what was going on deep down. Lavoisier, a stickler for going only where facts led him, had little time for invisible atoms.

Before atoms could be brought into the light of day, someone was going to have to build a bridge between their hidden realm and the world of chemistry. That person emerged from the wild and beautiful county of Cumberland in the north-west of England. His name was John Dalton.

IMAGINING ATOMS

John Dalton was born in 1766 in Eaglesfield, a small village surrounded by low rolling farmland in a remote part of north-west England. John's upbringing was decidedly modest; his father Joseph was a weaver by trade and the family owned and farmed a small strip of land near the village.

However, young John had a couple of advantages. First off, he was an unusually bright and precocious little boy, with a natural curiosity and the ability to soak up information like a sponge. Second, his family were Quakers, religious nonconformists who set a high value on learning. John's mother in particular encouraged his education and used the family's network among the Society of Friends to provide her son with a better schooling than a poor farm boy would normally have gotten in eighteenth-century England.

John developed an early fascination with the weather, which isn't surprising as there's a lot of it in the north-west of England. From his home he could watch rain clouds rolling in from the Irish Sea and passing over the dramatic peaks of Grasmoor and Grisedale Pike. The Quakers weren't exactly a fun-loving bunch – they were teetotallers and emphasized holy behaviour in all they did – but studying nature was one of the few permitted leisure activities, regarded as a way of revealing God's work in the world. As a boy John began to take daily readings of air pressure, temperature, humidity, and rainfall, a routine he followed until the day he died, and though he had no idea at the time, it was the beginning of a long journey that would eventually lead him to a theory of atoms.

Although John's education was supported by the Quakers, his situation was often precarious, and by the age of fifteen he was forced into agricultural labour to make ends meet. The future looked bleak, but salvation came with an invitation to teach at a Quaker boarding school fifty miles away in the market town of Kendal. The Quakers had generously equipped the school with a suite of scientific instruments that he was quick to start experimenting with.

He also acquired a much-loved tutor in the blind natural philosopher John Gough, who took a shine to the eager teenager and taught him mathematics and science, including Newtonian atomic theory. In return, John helped his blind mentor with reading, writing, and drawing diagrams for his scientific papers.

John had ambitions to study law or medicine but was barred from English universities because of his religion. Instead he eventually secured a position as a professor at a new college that had been set up by religious nonconformists in the booming industrial town of Manchester.

To the farm boy from Eaglesfield, Manchester was a huge and bustling place. Here, religious and political radicalism, new scientific ideas, and revolutionary technologies were driving change at a pace that was dizzying, perhaps even frightening. Manchester was the beating heart of an industrial revolution that was transforming Britain into the powerhouse of the world. Towering new cotton mills, powered by smoke-belching steam engines, and row upon row of redbrick terraced houses were rising on the city's skyline. Here science wasn't a hobby carried on by wealthy aristocrats in their private labs, but part of a thriving community of engineers, craftsmen, and industrialists. Dalton couldn't have come to a better place and dived headfirst into Manchester's larger scientific pond.

The weather remained his obsession, in particular, rain. It's a long-standing joke among southerners (like me) that it's always raining in Manchester. That may be a little unfair, but there is certainly no shortage of moisture in the north-west. Dalton would take long walking holidays in his much-loved but decidedly drizzly Lake District, where the air sometimes feels so heavy with water that you wonder if it could soak up any more. In fact, it was just this question that got him thinking about atoms.

Dalton began to do experiments to see how much water vapour a fixed volume of air could absorb. At the time, people thought that water dissolved in air, like sugar dissolves in a cup of coffee.

If you add more than around 150 teaspoons of sugar to a cup of coffee – which I think is even more than you get in a Starbucks cinnamon dolce latte – then it stops dissolving and you end up with sugar granules rolling around at the bottom of the cup. A similar thing happens when it rains: when the air is completely saturated with water vapour the water condenses into little droplets, which form clouds, and if the droplets get big enough it starts to rain.

However, if there is more air squeezed into a given volume then it should be able to soak up more water vapour. It's a bit like adding more coffee to your mug to dissolve those extra sugar granules. However, Dalton's experiments showed something truly weird: a container would always absorb the same amount of water vapour regardless of how much air was squeezed into it. It seemed as though the air and the water vapour somehow ignored each other, occupying the same space but without interacting.

What has all this got to do with atoms? I hear you cry. Well, it all comes down to the interpretation. Dalton took this result as evidence for the idea that air and water vapour only exert forces on atoms *of their own kind*. Two atoms of air would interact with each other, and two atoms of water vapour would interact with each other as well, but an atom of air and an atom of water vapour would totally ignore each other. It's a situation similar to the slightly awkward birthday parties I'd find myself at in my early twenties. There would usually be two groups: the birthday girl or boy's old schoolfriends and the newer university friends. Although we were all at the same party, we would drift around the room chatting within our respective cliques and barely acknowledge the existence of these *other* friends. According to Dalton, atoms of two different gases behave in more or less the same way.

Dalton published his theory in 1801, and it immediately caused a stir that spread beyond Manchester to the scientific academies of continental Europe. In London, the charismatic chemist and inhaler of strange gases Humphry Davy was intrigued by his theory of 'mixed gases', but many leading scientists argued passionately against it, including his old mentor and friend, John Gough, which must have stung a little.

Dalton was determined to prove his critics wrong and set out on a series of experiments that he hoped would provide irrefutable evidence for his theory. Along the way he became interested, almost by accident, in the problem of why certain gases dissolve in water more easily than others. His solution was simple but held the seeds of what would become a fully fledged atomic theory. Dalton argued that it was the weight of the atoms that determined how easily they dissolved, with heavier atoms dissolving more easily than light ones. To test this idea, he somehow had to figure out how heavy different atoms were compared to one another.

But how? Remember that no one had even got close to seeing an atom in the early nineteenth century. It would be almost two hundred years before a microscope would be invented that was powerful enough to image one. Atoms were just an idea and if they existed at all were so fantastically minuscule that almost every scientist of the day thought they would lie forever beyond our perception. How on earth could Dalton possibly measure their masses?

Dalton's stroke of brilliance was to take his theory of mixed gases – that atoms only repel other atoms of their own kind – and extrapolate it to figure out how many atoms of different chemical elements bind together to make molecules. His reasoning went something like this: imagine that two atoms of two different chemical elements, let's call them atom A and atom B, bind together to make a molecule A-B. Now imagine that another atom of A comes along and wants to join the party. Since atoms of A repel each other it will naturally want to get as far away from the other A atom as possible and so attach to the opposite side of the B atom to make a larger molecule A-B-A. Then if a third atom of A comes along this time it will arrange itself at 120 degrees from the other two atoms of A to form a triangular shape with B at the centre, and so on.

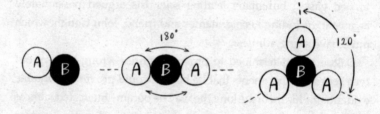

Dalton reasoned that if only one compound of A and B is known, then its molecule should have the simplest structure, which is AB. If there are two different compounds of A and B, then the second molecule will be the next simplest, ABA.

For example, two different gases made of carbon and oxygen were known in the early nineteenth century: one was called 'carbonic oxide' (the colourless toxic gas that nearly killed Humphry Davy when he breathed it in, possibly in the name of science or in search of another way to get high) and the other was called 'carbonic acid' (the fixed air discovered by Joseph Black, which was used to suffocate a number of unlucky mice, again in the name of science). By weighing the amount of oxygen that reacted with a fixed amount of carbon to make these two gases, Dalton found that carbonic acid contained twice as much oxygen as carbonic oxide. Applying the rules of his atomic theory, this meant that carbonic oxide was the simplest molecule, made of one carbon and one oxygen atom (what we now know as carbon monoxide, CO), and carbonic acid contained one carbon and two oxygen atoms (in modern terms, carbon dioxide, CO_2).

At last, Dalton could figure out the relative masses of carbon and oxygen atoms, calculating that an oxygen atom weighs about 1.30 times more than a carbon atom, which is remarkably close to the modern value of 1.33. Through a combination of guesswork, theorizing, and experimentation, he had measured a property of an atom, and in doing so had caught a glimpse of their hidden realm for the very first time.

Dalton knew that he was onto something big. He completely

forgot about the original problem of dissolving gases in water and engrossed himself in his new atomic theory. After three years of work, interrupted by heavy teaching duties and the occasional walking holiday in his beloved Lake District, he was ready to reveal his ideas to the world.

In March 1807 Dalton travelled to Edinburgh, arguably Britain's greatest intellectual and scientific centre and crucible of the Enlightenment. He was there to present what was nothing short of a revolutionary new description of the chemical elements. He began his momentous lecture series in the most English way imaginable, with an apology. 'It may appear somewhat like presumption in a stranger to intrude himself upon your notice in the character I am now assuming, in a city like this, so deservedly famous for its seminaries of physical science.' However, there was steel beneath Dalton's veneer of humility. He went on to announce that if the ideas he was about to share were borne out by experiment, which he was sure they would be, they would 'produce the most important changes in the system of chemistry, and reduce the whole to a science of great simplicity, and intelligible to the meanest understanding'.

The atomic theory Dalton presented at Edinburgh, and later published in his great work, *A New System of Chemical Philosophy*, finally connected Lavoisier's chemical elements with the ancient idea of atoms. According to Dalton, all matter was made up of solid, indivisible, indestructible atoms, and every chemical element was made of its own unique atom with a definite mass. Chemical reactions, from burning charcoal to baking apple pie, were nothing more than a process of rearranging these different atoms to make a wider variety of different molecules.

The response to Dalton's atomic theory was immediate, both in Edinburgh and beyond. In London, Humphry Davy was quick to see its potential to help chemists understand and quantify the way that different chemical elements reacted with one another. The theory's most important prediction was a rule known as 'the law of multiple proportions'. Basically, it says that when two elements react to make compounds, they always do so in certain ratios, which is a direct consequence of the fact that elements come in discrete little atomic lumps.

Take reactions between the two dominant gases in our atmosphere, nitrogen and oxygen, to make three different compounds: nitrous oxide, nitric oxide, and nitrogen dioxide. If we did three different experiments where we started with 7 grams of nitrogen and then reacted it with oxygen to make these three compounds, we would find that the amount of oxygen that joined up with the nitrogen in each case would be 4 grams, 8 grams, and 16 grams. From this, Dalton was able to figure out that the chemical formulae of nitrous oxide, nitric oxide, and nitrogen dioxide are N_2O, NO, and NO_2, and the reason that oxygen only reacts in these fixed proportions is because the mass of an oxygen atom is eight-sevenths the mass of a nitrogen atom.

Within months, other experimenters were finding evidence that the elements really did react in the way that Dalton's theory said they should, and soon Dalton was being feted around the country. In the same year that Dalton published his atomic theory, Humphry Davy tried to persuade him to become a fellow of the most prestigious scientific organization in Britain, the Royal Society in London.*

But while chemists were happy to take and apply the consequences of his atomic theory, far fewer agreed with Dalton's belief in real, physical atoms. In 1826, when Humphry Davy, now president of the Royal Society, presented Dalton with the Royal Medal,

---

* Dalton was having none of it and rejected Davy's advances out of hand. This proud northern radical had little time for the Royal Society, which he regarded as part of the corrupt political establishment. Its president, Joseph Banks, had stuffed the society full of his mates, and at the time the society was criticized as little more than a glorified gentlemen's club for dilettante aristocrats who dabbled in science. He only eventually joined in 1822 when some of his friends put his name forward without his knowledge.

he was keen to emphasize that this was for his work on the law of multiple proportions – a prediction of Dalton's atomic theory – and not for his belief in actual physical atoms.

Although Dalton had connected Lavoisier's chemistry with atomic theory, his ideas were too far ahead of their time. The debate over whether atoms existed or not was to rage for another hundred years and was only finally resolved by an aspiring young physicist working in the Bern patent office, whose destiny it was to change science forever.

EINSTEIN AND THE ATOM

You've got to feel sorry for Albert Einstein's schoolteachers. I mean, just imagine having *Albert Einstein* in your class. Of course, in 1895 his teachers didn't realize they were teaching *Albert Einstein*, just a puckish German teenager with a mop of unruly black hair and a self-satisfied smile.

Famously, Einstein was not a good student. At a fairly early age he had realized he could teach himself more advanced mathematics and physics than his teachers, and by his midteens had decided that school was a waste of time. He seems to have had a special talent for winding his teachers up. On one occasion his father Hermann was called into school to be berated for Albert's disruptive influence. When he asked what exactly his son had done, he was told by an exasperated teacher, 'He sits at the back and smiles.'

Despite a not entirely successful or happy schooling, Einstein was determined to pursue a career in physics and after one failed attempt got himself admitted to the Swiss Federal Polytechnic, a relatively new university in the city of Zurich. His time at the 'Poly', as it was known, was a happy one. He revelled in his newfound freedom and soon formed a tight-knit group of friends, spending most of his time in coffeehouses, sailing on the lake, or at parties entertaining groups of admiring young women with his violin playing. It was at one of these shindigs that he met his lifelong friend, Michele Besso, a mechanical engineer six years his senior, with whom he would spend many happy hours discussing the latest controversies in science, philosophy, or politics amid a haze of pipe smoke at their favourite cafe.

During one of their wide-ranging discussions, Besso introduced Einstein to the work of the Austrian physicist and philosopher Ernst Mach. Mach was an ardent opponent of atomic theory, arguing that atoms were little more than a convenient fiction that just happened to explain the behaviour of larger-scale objects. As long as atoms themselves remained out of the direct reach of human senses, Mach argued that belief in their existence was a matter of faith, not science.

Mach had a point. Almost a hundred years after Dalton published his chemical atomic theory, the evidence for atoms was still mostly circumstantial. That said, over the course of the nineteenth century, atomic theory had achieved several big wins. In chemistry, the marriage of atoms with chemical formulae (symbolic ways of representing different chemical compounds in terms of their atomic building blocks – such as N_2O for nitrous oxide) had proved extremely useful in exploring the reactions of organic molecules. Great progress had also been made on Dalton's project of measuring the relative weights of different atoms, resolving most of the ambiguities about the atomic make-up of molecules, including whether water was HO or H_2O.

At the same time a powerful new way of understanding the behaviour of gases had emerged, known as 'kinetic theory'. According to this theory, a gas was a multitude of minuscule atoms flying about through empty space, bouncing back and forth off the walls of their container like a swarm of tiny angry bees. This picture allowed physicists to neatly explain measurable properties of a gas like temperature and pressure. Lavoisier had thought of heat as a physical substance called 'caloric', which he included in his list of chemical elements. Kinetic theory did away with this idea; heat was simply a consequence of the speeds that the atoms were zipping about at. The faster the atoms were moving, the hotter the gas. This also explained why the pressure of a gas increases as you heat it up. As the temperature rises the atoms move about faster and hammer against the walls more often and with greater force, causing an increase in pressure.

An early version of kinetic theory had been proposed by Daniel Bernoulli way back in 1738 and had remained more or less unchanged until after the 1860s, when James Clerk Maxwell, Josiah Willard Gibbs, and Ludwig Boltzmann revamped the theory by applying statistics to describe how atoms continually bumping into one another determine the measurable properties of a gas. This new statistical theory was able to explain familiar phenomena like the conduction of heat or how long it takes for a smelly gas released on one side of a room to be noticed by people on the other,* as well as predicting wholly new ones.†

By the time Einstein was having his coffee- and tobacco-fuelled discussions with Besso in 1896, progress on kinetic theory had stalled. Despite its successes, the theory had gotten snagged on a couple of particularly thorny problems, leaving open the possibility that it could yet be overturned. But worst of all, it was still true that no one had seen an atom.

At the University of Vienna, a battle was raging for kinetic theory's very soul. On one side was Ludwig Boltzmann, the theory's leading light, and on the other, Ernst Mach, its arch-nemesis. Boltzmann was so stung by Mach's attacks that he devoted the last few years of his life to a valiant defence of his cherished kinetic

---

* Among other successes, kinetic theory provides a sound theoretical basis for the well-known rule that 'the one who smelt it dealt it'.

† Its greatest triumph was the completely counter-intuitive prediction that the viscosity or 'stickiness' of a gas doesn't increase as you increase the density of the gas, which was soon confirmed by experiment. This is really weird if you think about it; it implies that a pendulum swinging in ordinary air experiences no more resistance than a pendulum swinging in an airtight container with half the air pumped out.

theory, and although he won most physicists round to his side, Mach and a number of leading chemists remained intransigent.

In Zurich, the young Einstein followed the debates with increasing interest and frustration. He was convinced that Boltzmann was right and Mach was wrong. There was simply no way that all kinetic theory's successes were a fluke. Atoms were real, and as soon as he graduated, Einstein resolved that he would settle the two-thousand-year-old debate once and for all. Unfortunately, old habits die hard, and Einstein had not acquitted himself well in his studies, gaining the lowest passing grade in his year and earning a reputation as a 'lazy dog', as his favourite professor Herman Minkowski put it. He found himself struggling to find a job, eventually having to take up temporary teaching positions to make ends meet.

A reprieve came in 1902 when he got a job at the patent office in the Swiss city of Bern. Not only did this come with a salary that was twice what he would have received as a professor's assistant, but it was also undemanding enough that it allowed him to do scientific research on the side, both in his spare time and, as he later admitted, during working hours.

A steady income also made it possible for him to finally marry his university girlfriend, Mileva Marić. Mileva and Albert had met at the Poly (she had been the only female science student in his year) and formed an intense relationship that was both romantic and scientific. Einstein was clearly swept along by the prospect of a partner he could share both his life and his physics with and proposed marriage despite the opposition of his parents and the doubts of his close friends. Unfortunately, Mileva's ambitions for her own scientific career were thwarted when she failed her final exams – perhaps in part thanks to her boyfriend's bad influence – compounded by getting pregnant while she was doing retakes.

By 1903 the romance had clearly faded – Albert later said that he married her out of a sense of duty – but they nonetheless settled into a life of quiet domesticity. Mileva seems to have accepted the loss of a potential scientific career and the scandal of having a child out of wedlock with remarkable stoicism, cheerfully taking care of the home and more or less all of her husband's needs. Combined with his light duties at the patent office, this trouble-free life set Einstein up for the most productive period of his entire career.

The year 1905 has mythic status in the history of science. Over a period of just a few months, Einstein published four papers, each of which sent shockwaves through the physics world that are still being felt today. Two of the four were truly revolutionary: one upended the fundamental concepts of space and time, the other heralded the dawn of the quantum age. Relativity and quantum mechanics – two beautiful, deeply unsettling ideas that challenge our most basic notions of how the world ought to work – are the pillars upon which modern particle physics is built. (We'll come back to them again and again in the coming chapters, but we're not quite ready to discuss them yet.)

It is incredible that the paper that finally proved the existence of atoms was arguably the least revolutionary of the four. There is a reason that 1905 is referred to as Einstein's 'miraculous year'. Einstein's warm-up act was his PhD dissertation on what sounds like the rather curious subject of sugar solutions but was actually an ingenious method to calculate the number and sizes of sugar molecules. Even though Einstein got a result that was remarkably close to the accepted modern values, this still didn't constitute proof of the existence of molecules or atoms – his calculations were based on the same bunch of unproven assumptions that formed the basis of kinetic theory.

What Einstein needed was a smoking gun, an unmistakable signature that could only be left by an atom. He knew that atoms were far too small to be seen directly through a microscope, but what if there was a way of seeing their influence on particles that *were* large enough to be visible?

In 1827, the Scottish botanist Robert Brown had discovered a peculiar phenomenon when peering through his microscope at some pollen grains. Within the grains, he noticed tiny particles endlessly jiggling about. While many suggestions had been made to account for the effect, including living molecules within the pollen and vibrations from passing carriages, no good explanation for the jiggling, which became known as 'Brownian motion', could be found at the time. Three decades later in the 1860s, a couple of scientists had suggested a new explanation: What if the pollen particles were moving about thanks to taking continuous blows from individual water molecules? The water molecules themselves might be far too small to see through a microscope, but perhaps their influence could be seen each time they crashed into a much larger particle. The problem is that a single water molecule is far too small and moves far too slowly to have any noticeable effect on the position of a relatively ginormous pollen particle. It would be the equivalent of an aircraft carrier being noticeably deflected by a collision with an anchovy.

Einstein realized that even though an individual water molecule couldn't visibly move something as large as a pollen particle, the accumulated effect of a large number of collisions might. According to kinetic theory, a pollen particle floating in water is surrounded by thousands of water molecules, which are all jittering about thanks to the heat of the water. Because of the inherent randomness of this jittering, sometimes one side of the pollen particle will get hit by more water molecules than the other, creating a net force large enough to make it move.

This cumulative effect causes the pollen particle to follow what is known as a 'random walk' through the liquid, a zigzagging path that looks a bit like a drunkard stumbling around in the dark. At one moment the pollen particle will be pushed in one direction, and then a moment later in a different, random direction. Even though each step on this journey is random, over time the particle gradually moves farther and farther away from its starting point. Einstein's aim was to connect the average distance that a pollen particle moves in some fixed amount of time to the number of molecules in a given volume of water.

After some brilliant physical insights and very clever mathematics, he arrived at a single equation that says that the distance a pollen particle jiggles away from its starting point in some amount of time goes up as the number of water molecules goes down. Now let's think about the big argument that Einstein was trying to settle: one side said that matter is made of atoms, the other that atoms are just a figment of physicists' imagination and that matter is continuous. If matter is continuous, then that means you can divide up any object, be it an apple pie or a drop of water, into an infinite number of infinitely small bits. Or to put it another way, there are an infinite number of infinitely small water molecules in a drop of water. If that were true, then according to Einstein's equation, a pollen particle wouldn't move at all, which sort of makes sense if you think about it. If the number of water molecules is effectively infinite then there will always be an equal number (that is, infinity) pushing on the pollen particle from any given direction, which means that the forces experienced by the particle are always perfectly balanced and hence the pollen particle stays dead still.

But the pollen particles do move! In other words, Einstein had shown that you could *only* explain Brownian motion if atoms were real. Not only that, but he had provided a new way to calculate the number of water molecules in a drop of water based on how far a pollen particle wanders in a given amount of time.

Now this all sounds very neat and tidy, but unfortunately the history of science is never quite as straightforward as that. Einstein didn't actually set out to explain Brownian motion. His aim was to find a way to prove atoms existed, and it seems that it was only after he'd done his calculations that he realized there might be a link with Brown's jiggling pollen particles. To seal the deal, Einstein needed experimental proof that the way small particles wander about in the water corresponds precisely to his equation. At the end of his paper he threw down the gauntlet to his experimental colleagues: 'It is to be hoped that some enquirer may succeed shortly in solving the problem suggested here, which is so important in connection with the theory of Heat [kinetic theory].'

It was the French physicist Jean Baptiste Perrin who eventually took up Einstein's challenge. Between 1908 and 1911 he and his team of research students performed a series of tour-de-force experiments that verified Einstein's predictions in every way. Einstein's theoretical brilliance and Perrin's experimental guile had finally proven old John Dalton right. The age-old debate was finally settled. Matter is made of atoms.

At last we can answer Carl Sagan's original question: how many times do you need to cut an apple pie in half until you get down to an individual atom? Along with verifying Einstein's equation, Perrin also measured Avogadro's number, which allows you to figure out the number of atoms or molecules in a given mass of a substance, including, for instance, an apple pie. Popping one of Mr Kipling's finest on my kitchen weighing scales and doing a quick calculation reveals that a single apple pie contains roughly four trillion trillion atoms!

How many times would we need to cut the pie in half to get down to one of these atoms? Well, in *Cosmos* Sagan tells us that the answer is twenty-nine. His apple pie was a bit bigger than mine, though, so I thought I'd better check this for myself. After running the numbers, I was appalled to find that the great Carl Sagan had got it wrong! His calculation assumes only cutting the pie in one dimension, which would give you a slice one atom thick but as high and deep as the original pie. The correct approach is to ask how many cuts do we need to make until the last two pieces are each a one-quarter of a trillionth of a trillionth share of the initial apple pie. In other words, one atom. This gives you the correct answer of eighty-two cuts. There's a letter of correction winging its way to the producers at PBS as we speak. Sorry, Carl.

Anyway, a good scientist should test their theoretical predictions, so I grabbed my best kitchen knife and had a go. After about four-teen cuts I was left with a crumbly mess and, I confess, none the wiser about the atomic structure of apple pie. The problem is that atoms are just too fantastically small: a single carbon atom is around a tenth of a billionth of a metre across. If you struggle to imagine something that tiny, then an analogy from the great theoretical physicist Richard Feynman may help. If you took an ordinary apple and blew it up to the size of the Earth, then one of its atoms would end up around the same size as the original apple. No knife made by humans is capable of slicing an apple pie finely enough to get down to something that small. How then could I check if the apple pie really is made of atoms? Actually, all you need is a pestle and mortar and a microscope.

First off, I ground up some of the black apple pie charcoal that we'd made in the first experiment. Unfortunately, it turned out that my charcoal wasn't as pure as I'd thought; it must still have contained quite a lot of oils and moisture and formed a paste instead of the fine dust I was after. After some vigorous heating to drive off the last impurities I managed to get the dry powder I wanted. I proceeded to place a small drop of the yellowish apple pie liquid onto a microscope slide, dusted it with a tiny amount of the char-coal, slid the slide onto the microscope's stage, and peered downward.

At four-hundred-times magnification the powder particles were huge, filling almost the entire field of view. I was worried that I'd not ground the charcoal up finely enough and was about to remove the slide when I noticed a group of much smaller black particles in the bottom left. Letting my eye adjust and keeping as still as possible, I suddenly saw it. They were moving. Not in a gentle flowing way that might have suggested currents in the liquid, but with an agitated jittering. I could see immediately why Brown had initially thought he'd discovered living molecules; they really did look as though they were dancing about. I was genuinely delighted, a similar feeling to when I first looked through a telescope at a yellowish dot in the sky to see a perfect little image of Saturn, complete with rings and pinprick moons, hanging in the blackness of space. It may sound silly, but on seeing Saturn my instant reaction had been 'Oh, my God, it's real!' Images in books or on TV are one thing, but seeing it with my own eyes made it real to me in a way it had never been before.

Those dancing black specks of incinerated apple pie had a similar, and completely unexpected, effect on me. To think that each wiggle, zig, and zag was caused by untold numbers of unseen blows from indescribably small and yet (suddenly) undeniably physical atoms was strangely affecting. As a physicist, the concept of atoms is so familiar it can breed a kind of unthinking complacency, and I realized that this was one of the few times I had really seen evidence of their existence with my own eyes, proof positive that at least some of this particular apple pie really was made of atoms.*

Of course, atoms are not the end of the story. Paradoxically, signs that they were made of even smaller things had been being uncovered in the labs of Europe for at least a decade by the time Perrin's experiments sealed the deal for their existence. The consequences of these discoveries were to prove profound, triggering a revolution in our understanding of matter and the laws of nature, while unleashing forces that had been hitherto unimaginable.

---

* Strictly speaking, it proves that the acrid yellow liquid that came off the apple pie is made of molecules, since it's the molecules in the liquid that continually hammer against the black particles and cause the jittering motion.

CHAPTER 3

The Ingredients of Atoms

Atoms are small. Stupendously, indescribably, unimaginably small. How small? Well, you could line up around 5 million carbon atoms across the full stop at the end of this sentence. That probably doesn't really help, to be honest. It's a pretty hard ask to picture something that's less than a millionth of a millimetre across. After all, what's the smallest thing you've ever seen with your own eyes? Perhaps a speck of dust floating in a sunbeam, or a flea. Well, they're both bloody gigantic next to an atom.

Given their mind-boggling smallness it's pretty astounding that we can say anything at all about what atoms themselves are made from. That we can is ultimately down to four brilliant and, by today's standards at least, staggeringly simple experiments carried out over a few decades around the start of the twentieth century.

This was the heroic age of experimental physics, when really profound discoveries could be made by just one or two people, beavering away in a dingy university laboratory. Today, making a major breakthrough in particle physics requires a gargantuan international effort involving thousands of physicists, engineers, and technicians and millions, if not billions, of euros, dollars, pounds, and yen. I work on the LHCb experiment with more than 1,200 people from across the globe, and we're actually the smallest of the four big detectors at the Large Hadron Collider, a machine that itself took almost four decades to plan, design, and build. On the other hand, the very first subatomic particles were all found using equipment cobbled together on a shoestring budget that could fit comfortably on a laboratory workbench.

So as we probe deep into the atom in search of the basic ingredients of our apple pie, I'd like to take you back to this heroic age when the basic ingredients of atoms were first discovered. But before we do, it's worth recapping what was thought about the structure of matter at the end of the nineteenth century. From John Dalton we have the idea that every chemical element is made from a corresponding atom, the smallest possible unit of matter. However, the indivisibility of the atom wasn't universally accepted. In 1815, the English chemist William Prout had argued that all the different elements might ultimately be made out of hydrogen atoms stuck together, which he based on the curious fact that all the elements appeared to have atomic masses that were roughly whole number multiples of hydrogen's. However, Prout's hypothesis wasn't widely accepted, partly since he'd obtained his data from some distinctly dodgy experiments combined with a bit of judicious rounding: partly because of awkward elements like chlorine, whose atomic mass was 35.5 hydrogens; and also because many chemists were appalled by the prospect of resurrecting the alchemists' old get-rich-quick scheme of converting lead into gold, which if Prout was right, became a simple matter of chipping a few hydrogens off a lead atom.

The other big piece of circumstantial evidence in favour of atomic substructure had come in 1869, courtesy of the Russian chemist, inspector of cheese factories, and hairdressers' despair* Dmitri Ivanovich Mendeleev. After several long train journeys spent playing a game of chemical solitaire with cards representing the

---

* He insisted on getting his hair cut and beard trimmed only once a year, cultivating a look that was a striking blend of Gandalf the Grey, Leonardo da Vinci, and Fagin.

different elements, Mendeleev noticed that when he ranked the elements by atomic weight, their chemical properties repeated with a peculiar regularity. Laying out the elements to form the periodic table, Mendeleev was able to predict the existence of three brand-new elements, which he believed were required to fill the gaps in his scheme. Within a few years, they obligingly turned up – gallium, scandium, and germanium – with more or less the precise properties that he had predicted.

Where did these relationships between the chemical elements come from? At the very least, the periodic table showed conclusively that the elements were not a random collection of unrelated ingredients. There was clearly some sort of deeper order to the properties of atoms, and while that didn't necessarily imply substructure, it was a tantalizing hint. However, the spectre of alchemy loomed so large that it would take powerful experimental evidence to persuade chemists and physicists that atoms were really made of smaller things. Which brings us to heroic experiment number one, carried out in the dusty Cambridge laboratory where particle physics was born.

PLUM PUDDING

On a sleepy lane, nestled out of sight behind Cambridge's Corpus Christi College, stands what should be one of the most famous buildings in the world, the original Cavendish Laboratory. Just a stone's throw away on the bustling King's Parade, crowds of tourists, boat tour touts, aggravated taxi drivers, and bicycle-riding students jostle for space, but here all is usually serene. Few on the tourist trail make it to the Cavendish, instead spending their time gawping at Cambridge's medieval architecture or being taken on exorbitant punt rides up the river. But, every so often, you'll see a small group gather outside the old lab, sometimes sheltering from the English drizzle under its arched entrance, while a guide shoots off a quick-fire list of world-changing discoveries that were made within its walls. Then after maybe five minutes, they shuffle off, normally in the direction of the Eagle pub, where Cavendish researchers Francis Crick and James Watson famously announced that they'd discovered 'the secret of life' in the double-helix structure of DNA.

Aside from a small plaque fixed to its front wall, there's precious little evidence that anything of much significance happened here, which is a source of eternal frustration to me. If particle physics were a religion, this would be its holiest of holies. Hordes of pilgrims would throng to the old Cavendish every year to walk the corridors and touch the stones where men and women once split atoms and uncovered new building blocks of nature. Perhaps there'd be a gift shop selling kitsch porcelain figurines of Ernest Rutherford and J. J. Thomson. Anyway, particle physics isn't a religion, which is probably for the best, so instead when the physics department abandoned the creaking Victorian building for more spacious digs on the edge of the city in the mid-1970s, the university filled the old lab with social scientists and whacked up a plaque as a sop to history.

That hasn't stopped the odd pilgrim making their way here over the years. Just after the apple pie scene in episode 9 of *Cosmos*, Carl Sagan himself turns up in the Cavendish's old lecture theatre, which he declares to be the place where 'the nature of the atom was first understood'. That's a bit of an exaggeration as we'll see, but Cavendish physicists certainly had a claim to a large part of the puzzle, the first piece of which was found in the dying years of the nineteenth century, by the lab's top prof, Joseph John Thomson.

'J.J.', as he was known affectionately to his students, had been a bit of an odd choice to head up one of Britain's leading experimental laboratories. He was a mathematical physicist by training and notoriously clumsy, to the extent that his laboratory assistant often did his best to keep his boss from handling the delicate glass bulbs that they used in their work. However, Thomson did have a knack for designing ingenious experiments and a keen nose for an interesting problem, one of which arrived like a bolt from the blue at the start of 1896. In Germany, Wilhelm Röntgen had discovered a miraculous new type of ray that could penetrate human flesh and reveal the bones beneath. He caused a worldwide sensation when he shared a macabre photograph of the bones of his wife Anna's hand; horrified by the image, she remarked, 'I have seen my death.' The mysterious new form of radiation was given a suitably enigmatic name: the X-ray.

Röntgen had found X-rays emanating from the end of a Crookes tube, a glass bulb with most of the air pumped out and two electrodes inside. It had been known for several decades that when a high voltage was connected to the tube, so-called cathode rays would flow from the negative electrode (the cathode) towards the positive electrode (the anode), creating an eerie green glow where they hit the end of the tube. Röntgen's X-rays appeared to be coming from the place where the cathode rays struck the glass, but despite having been known about since the 1860s, no one was sure what cathode rays actually were.

Inspired by Röntgen's discovery and the scientific potential of the new X-rays, Thomson set himself the task of uncovering the true nature of cathode rays. At the time there were broadly two schools of thought: either they were some form of electromagnetic wave, like radio, light, or the new X-rays, or they were a flow of negatively charged particles, most likely electrically charged atoms known as ions. Thomson, who had spent most of the past few years breaking gases apart into ions using electricity, was firmly of the latter view. The question was, how to prove it?

In 1895, Jean Baptiste Perrin had shown that if you fired a beam of cathode rays into a metal cup, a negative electric charge built up, which he took as evidence that they were indeed negatively charged particles. However, many physicists weren't convinced, arguing that the negative electric charge might just be a side effect of the cathode rays, rather than an intrinsic part of them.

Picking up where Perrin had left off, J.J. carried out a modified version of the experiment, but this time he placed the cup at an angle, out of the firing line of the cathode rays. When the tube was switched on, the rays travelled in a straight line, missed the cup, and no negative charge built up. However, when Thomson used a magnetic field to bend the cathode rays off their straight path and into the cup, hey presto! Negative charge was detected. In other words, the electric charge seemed to go wherever the cathode rays did. If he had had any doubts before, J.J. was now absolutely convinced that cathode rays were negatively charged particles, but what type of particle was still unclear. Were they atoms, or something else entirely?

To figure it out once and for all, J.J. needed to know the mass of a cathode ray. If his guess that they were negatively charged atoms was correct then he'd expect them to have a larger mass than the lightest known atom, which was hydrogen. The tricky part was how to measure the mass of something so fantastically tiny. Remember that we're still more than a decade away from being able to measure the size of an atom, even indirectly.

However, there was a way to do it: to look at how the particles curved when they passed through a magnetic field. The heavier the particles were, the less they would bend (think about a truck driving round a corner: the heavier the truck's load, the more friction needed in the tyres to keep it from sliding off the road). The trouble was that the amount that the particles bend also depends on their speed and electric charge. A fast-moving particle bends less than a slow-moving one, while the bigger the electric charge carried by the particle, the stronger the magnetic force it feels, and the more it gets bent. As a result, J.J. couldn't measure their masses directly. He could, however, compare their mass to their electric charge.

When J.J. did his calculations he found something shocking. The mass of the cathode ray divided by its electric charge was roughly a thousand times *smaller* than a hydrogen ion's. There were only two possible interpretations: either the electric charge of a cathode ray was much bigger than a hydrogen ion's, or its mass was smaller, perhaps thousands of times smaller. Could Thomson have glimpsed something even more fundamental than Dalton's indestructible atom?

On Friday, 30 April 1897, J.J. caught the train down to London to present his findings at one of the Royal Institution's famous Friday Evening Discourses. The steeply ranked rows of the oak-panelled theatre were stuffed with the great and the good of the scientific establishment, dressed formally in their evening wear. Standing behind the desk where Humphry Davy had once captivated audiences with his dramatic chemical experiments, Thomson made his case for a new theory of the atom. Speaking with a slight Lancashire lilt and shuffling about the lecture theatre in his own inimitable way, he guided his audience through his recent experiments, laying out the evidence that cathode rays were tiny negatively charged particles, far smaller than the smallest atom. That might have been a startling enough claim on its own, but he was building up to his final, radical conclusion. These particles, which Thomson named 'corpuscles',* were the basic building blocks of all atoms. The electric forces inside his glass tubes were literally tearing atoms apart, setting streams of negatively charged particles free from their atomic prisons. He had divided the indivisible atom!

There was general disbelief among the scientists in the audience. One witness later said he thought that Thomson was 'pulling their legs'. While many were happy to accept his claim that cathode rays were negatively charged particles, the idea that the rays were the constituents of atoms was a step too far. Thomson had clearly got carried away, going far beyond what was supported by his experiments. If he was going to convince the doubters of his extraordinary claim, he was going to need really extraordinary evidence.

---

* The same term used by Isaac Newton to describe his hypothetical particles of light.

Back in the lab, Thomson enlisted his laboratory assistant Ebenezer Everett, described as 'the best glassblower in England', to handcraft him a cathode ray tube of such strength and precision that it would let him settle the debate once and for all. The new tube needed to have additional electrodes bled through the glass so that an extra electric field could be applied to the cathode rays, and it had to be able to withstand incredibly high vacuums, since almost every last trace of air would have to be pumped out of the vessel for the experiment to work. That unenviable task fell to Everett, who spent several days laboriously pumping the tube out by hand.

The pair worked hard over the summer of 1897. Everett, who wasn't about to let his cack-handed boss anywhere near his beautifully blown glassware, did most of the hands-on stuff, while Thomson was kept at a safe distance and only allowed in close to take readings. By comparing how magnetic and electric fields of different strengths deflected the cathode rays, Thomson arrived at a far more precise measurement of the mass-to-charge ratio, in perfect accordance with his earlier results. Their hard work had paid off; Thomson's corpuscles really did seem to have a mass thousands of times smaller than hydrogen.

In October that year he released a new paper, reaffirming his bold claim that corpuscles were the building blocks of atoms. But this time he went even further, laying out a model of the atom in which the corpuscles arranged themselves in concentric rings in a sea of positive electric charge. Over a number of years, he developed his picture of the atom, taking inspiration – rather appropriately given the title of this book – from a popular English dessert of the period. According to Thomson the atom was like a plum pudding, with negatively charged corpuscles acting as the plums, embedded in a positively charged sponge cake. At last Mendeleev's periodic table could begin to be unravelled; the properties of the different chemical elements were the result of them containing different numbers of corpuscles.

It took several years for Thomson's ideas to be accepted. The spectre of alchemy still haunted the physics community, with many unwilling to countenance the idea of a *subatomic* particle. One thing that never stuck was J.J.'s name for his particle – you've probably never heard of a corpuscle, and that's because we now know them as electrons. To this day, every experiment that has ever been carried out suggests that electrons are truly fundamental objects.

We have arrived at the first true ingredient of our apple pie.

But the story of the atom is far from over. While J.J. had been playing around with electrons, one of his young students, only recently arrived from New Zealand, had begun a journey that would propel him and the entire field of physics into a brave new world. His name was Ernest Rutherford, and he was to change how we think about atoms forever.

THE HEART OF THE ATOM

Much as I'm proud of my home university, Carl Sagan was definitely giving Cambridge too much credit when he claimed that it was the place where the atom was first understood. In reality, it was in the forward-thinking industrial city of Manchester where the modern atom was forged. For more than a decade, a tight-knit band of physicists at the university's physics laboratory unravelled the secrets of the atom, under the leadership of, in my view, the greatest experimental physicist of all time, Ernest Rutherford.

Rutherford, a farmer's son from Pungarehu on New Zealand's North Island, had come to Britain in 1895 as one of J. J. Thomson's research students at the Cavendish. He quickly made a name for himself as a brilliant experimenter, but it was only towards the end of his time there that he picked up the trail that would eventually lead him to uncover the true structure of the atom. In 1896, Henri Becquerel in Paris had discovered a new form of radiation that sprang spontaneously from minerals containing uranium, and

Rutherford made the risky decision to drop his promising work on X-rays and throw himself headfirst into studying the mysterious phenomenon. It was a decision that would prove to be the making of him.

In 1898 he had left Britain to head up the physics laboratory at McGill University in Montreal, Canada, at the tender age of twenty-seven, soon transforming the lab into one of the two great centres for radioactive research. The other was in Paris, where Marie Curie and her husband Pierre carried out a series of gruelling experiments that involved stirring steaming vats of pitchblende (a uranium-rich mineral) in the open air until they had painstakingly extracted a few tenths of a gram of an element that was millions of times more radioactive than uranium. They named it 'radium'.

With the Curies and Rutherford leading the charge, order was gradually imposed on the growing list of radioactive elements. One thing slowly became clear to both Ernest and Marie: radio-activity must come from somewhere inside the atom itself, and what's more it seemed to change the original atom into a completely different one in the process. Working with the chemist Frederick Soddy at McGill, Rutherford amassed incontrovertible evidence that the radioactive element thorium decayed into a second element, which they dubbed 'thorium-X' (now known to be radium), which in turn decayed into a radioactive gas. For the first time in history they had caught a supposedly immutable element transmuting into a totally different one. The alchemists seemed to be back in business.

In 1907, Rutherford had been lured back to England by the prospect of being closer to the thick of the scientific action in Europe, much to the dismay of his colleagues at McGill. He arrived in Manchester like a whirlwind, reshaping the lab into something altogether new to science, a research school where almost all work was focused on what he now regarded as the most important problem in physics: the inner workings of the atom. Under his leadership, the number of staff and research students swelled, with

Rutherford dishing out projects designed to attack the subject from every possible angle. Time and success had changed him from a slightly shy, if determined, young man, into a boisterous, self-confident, and inspiring leader, a larger-than-life figure who one colleague compared to a living lump of radium. His booming voice would announce his approach long before he became visible as he made his daily rounds of the lab while tunelessly bellowing 'Onward, Christian Soldiers', dropping in on his staff and students to discuss whatever problem they were grappling with and dole out advice.

That said, working with Rutherford wasn't always easy. He had a volcanic temper, which would erupt without warning and over-whelm anyone unfortunate enough to be nearby at the time. Shortly after arriving in Manchester, he publicly dressed down the professor of chemistry, who had been slowly encroaching on lab space reserved for physics, bringing his fist down on a desk and exclaiming, 'By thunder!' before pursuing the unfortunate prof to his office, shouting that he was 'like the fag end of a bad dream'. His rages could be terrifying, as he loudly excoriated his victim from close range, although after the red mist cleared, he would almost always return to make a somewhat shamefaced apology.

Despite his volatility, Rutherford was loved and revered by his team at Manchester. They were more than just a group of scientists; they were a family, bound together by a sense of being at the fore-front of the most important scientific work being done anywhere in the world. Rutherford had an uncanny knack for choosing the right phenomenon for investigation, boasting that he never gave a student a dud project. Perhaps his greatest quality was his sheer doggedness, relentlessly hammering away at a problem until it gave up its secrets. One of his longest-serving colleagues, James Chadwick, was once asked if Rutherford had an acute mind. He replied that 'acute' was the wrong word. 'His mind was like the bow of a battleship. There was so much weight behind it, it had no need to be as sharp as a razor.'

The problem that was now in Rutherford's crosshairs was the structure of the atom itself. Despite all the progress in radioactivity over the past decade, many mysteries remained. No one knew why some atoms suddenly decided to change into others and spit out radiation. Even more mysterious was where the energy released in radioactivity came from. Rutherford had calculated that the radioactive decay of an atom released millions of times more energy than the most violent chemical reaction. There must be some vast well of energy deep within the atom itself, but what it could be was anyone's guess.

He hoped that he might be able to find answers by focusing on the particles that came flying out when a radioactive atom decayed. While still a student at the Cavendish, Rutherford had discovered that there were actually two different types of radiation being shot out by uranium: one that only flew a few centimetres through the air before stopping, and a second more penetrating ray that could travel much farther and even pass through strips of metal. Rutherford named these two types of radiation after the first letters of the Greek alphabet: alpha and beta.* Scientists had been quick to find that the highly penetrating beta rays could be bent using a magnetic field and soon realized that they were electrons.

Rutherford's first big success at Manchester was proving what he had long suspected, that alpha particles were helium atoms that had lost two electrons. The work had been done with a promising young German physicist, Hans Geiger, who pulled off the remarkable feat of building the first detector capable of counting alpha particles one by one.† While Geiger and Rutherford had been perfecting their detector, they had been irritated to find that the images left by a beam of alpha particles on a photographic plate

---

* A third type of even more penetrating ray, named 'gamma', was described by Paul Villard in 1900.
† This detector was the forerunner of the modern Geiger counter, which is still used today to measure radiation levels, emitting an ominous clicking sound beloved by directors of disaster movies.

became fuzzy after they had passed through the long tube of gas that made up the detector. This seemed to suggest that the alpha particles were being knocked off course by collisions with the gas molecules. Rutherford was puzzled; alpha particles got shot out of disintegrating atoms at incredibly high speeds, zipping along at a decent fraction of the speed of light. How could projectiles of such 'exceptional violence', as he put it, be deflected by something as insubstantial as a gas molecule?

Once again, Rutherford had shown his uncanny ability to ask precisely the right question. He set Geiger the task of firing alpha particles through a range of different materials and measuring how much they got scattered. After trying out thin foils of various metals, Geiger found that the heavier the atoms in the metal foil, the more the alpha particles appeared to be deflected. Gold turned out to be the best deflector of all, often scattering the alpha particles through such large angles that it left Geiger and Rutherford scratching their heads.

Why was this all so surprising? Well, according to J. J. Thomson's plum pudding model, the atom was a weak and wobbly sphere of positive charge (the sponge cake), with tiny negatively charged electrons (the plums) stuck into it. It was hard to imagine how something so diffuse and insubstantial as an atom could cause any trouble to something as powerful and speedy as an alpha particle.

It was while Rutherford was mulling over these peculiar results that he made an off-the-cuff suggestion: that one of the new students, Ernest Marsden, look to see if any alpha particles bounced backwards off the gold foil. Rutherford was sure that Marsden wouldn't see anything – there was absolutely no way a gold atom could knock an alpha particle backwards – but it would be a good project to get him trained up in radioactive research.

Counting alpha particles in those days was punishing work. I sometimes feel like a fraud when I describe myself as an experimental physicist – in truth I rarely come close to having to do anything hands-on. Data from the Large Hadron Collider is delivered via the Internet straight to me in the comfort of my Cambridge office, an airport departure lounge, or even while sitting in bed with a nice cup of tea. At the start of the twentieth century, on the other hand, you would have to sit in a darkened room, peering through a microscope at a zinc-sulphide screen for hours on end, patiently and methodically counting faint flickers of light that were the telltale signs of alpha particles, until tired eyes forced you to call it a day. And all this with your head just a few centimetres from a powerful radioactive source.

Before he began to take measurements, Marsden sat alone in the darkened lab for twenty minutes, slowly allowing his eyes to adjust to the gloom. On the workbench in front of him was a delicate glass cone containing the source of the alpha particles, a highly radioactive mix of radium, bismuth, and radon gas, with a thin mica window at one end to let the alpha particles escape. In their path hung the target, a thin golden foil that glimmered gently in the dim light of an electric lamp. On the same side of the foil as the radioactive source but shielded from it by a lead barrier to stop alpha particles from hitting it directly, was the zinc-sulphide screen and the microscope.

Once his eyes had fully adjusted, Marsden leaned in and put one eye to the microscope. He knew that Rutherford hadn't expected him to see anything at all, but he was astounded to see that the screen was alive with tiny flickers of light. They arrived sporadically, like the flashes of dozens of infinitesimal cameras at some microscopic movie premiere. Fearing Rutherford's ire, the young undergraduate student checked and rechecked his results until he was absolutely convinced that he hadn't messed anything up. After three days of eye-straining work, he finally gave Rutherford the jaw-dropping news as he was coming down the stairs from his study at the top of the laboratory: the alpha particles were being knocked backwards!

Rutherford was stunned. He later described it as 'quite the most incredible event that ever happened to me in my life. It was almost as incredible as if you fired a 15-inch shell at a piece of tissue paper and it came back and hit you.' Neither Geiger nor Marsden had the foggiest idea what was going on. When they published their extraordinary results in July 1909, they didn't even try to explain what they'd seen, except to make the tantalizing observation that it would require a magnetic field billions of times stronger than was possible in the lab to bend alpha particles back on themselves. Whatever it was that was sending the alpha particles hurtling backwards had to be the seat of almost unimaginably powerful forces.

Even Rutherford was stumped. Having produced an astonishing fourteen scientific papers in 1908, his research slowed to a trickle as he brooded over the puzzle. He took to spending long periods deep in thought, locked away in his study at home, endlessly turning the problem over in his mind. At first, he wondered whether the alpha particles that came bouncing back had actually banged into multiple gold atoms, with dozens of tiny knocks building up until the alpha particle was turned back on its heels. But his calculations showed that the chances of that happening were vanishingly small, far too low to explain the number of flickers that Marsden had seen. The alpha particles had to be being reflected in a *single* collision with something with a lot of mass.

It was over a weekend in December 1910, more than eighteen months after Marsden had given him the startling news, that Rutherford finally saw the answer clearly. Charles Galton Darwin,* a young student at the Manchester lab, had been invited round to the Rutherfords' for Sunday dinner, and after they'd finished eating, Rutherford shared his world-changing insight for the first time. His old mentor J.J. had it all wrong; the atom wasn't a pudding-like blob, it was a tiny solar system, with negatively charged electrons

---

* Grandson of the famous naturalist Charles Darwin, author of the theory of evolution by natural selection.

held in orbit around an infinitesimal positively charged sun,* hidden deep at the heart of the atom. This atomic core, which Rutherford would later name the nucleus, contained 99.98 per cent of the atom's mass crushed into a tiny speck 30,000 times smaller than the atom itself. It was this tiny but mighty nucleus that was responsible for scattering the alpha particles back on themselves. On the rare occasion that a positively charged alpha particle came close to the nucleus, it experienced an incredibly powerful electric repulsion, and if the collision was more or less head-on, the repulsion would send the alpha particle zooming backwards.

The next morning a triumphant Rutherford bounded into the lab, his face beaming, and went straight to tell Geiger that he finally knew what the atom looked like. That very day Geiger got going on experiments to test the rough predictions of Rutherford's model. After a few more weeks of bombarding gold foils with alpha particles, Geiger found that the angles they came ricocheting off at agreed neatly with Rutherford's predictions. In March 1911 Rutherford felt ready to show his atom to the world. Appropriately, he chose the Manchester Literary and Philosophical Society, where John Dalton had discussed his own atomic theory a century earlier.

What Rutherford revealed that day was more than just the discovery of the atomic nucleus; he had seen, for the very first time, the shape of the subatomic world. The nucleus contains almost all of an atom's mass squeezed into a tiny space, tens of thousands of times smaller than the atom itself, orbited by insubstantial clouds of electrons. If you were to blow up an atom to the size of something more familiar, say a football stadium, the nucleus would be about the size of a marble sitting at the centre of the field, with the electrons whizzing around somewhere up in the stands.

However, there was a serious problem with Rutherford's atom. If the atom was really like a tiny solar system, then it couldn't

---

* Actually, at the time Rutherford wasn't sure whether the centre of the atom had a positive or negative charge – this was only decided on a few years later.

possibly be stable. It's a well-established law that when a charged particle is accelerated, in a circle for instance, it gives off electromagnetic radiation. This implied that the electrons should be constantly emitting light, losing a bit more energy with each orbit, until they eventually spiralled into the nucleus. Previous attempts to propose solar-system-like atoms had failed for this very reason. J.J.'s plum-pudding metaphor for the atom had been largely motivated by an attempt to find a theoretical arrangement of electrons that was stable against collapse.

The solution to this paradox came from a brilliant young Danish physicist named Niels Bohr, who enlisted a strange new idea known as the 'quantum'. In the early twentieth century, Albert Einstein and Max Planck had put forward the idea that light comes in discrete little lumps, or quanta. Taking inspiration from this idea, Bohr argued that electrons could only move around the nucleus in certain fixed orbits, emitting quanta of light as they jumped from one level to another. And as the electrons were restricted to only these levels, like trains moving on circular tracks, it was impossible for them to fall into the nucleus. Bohr's marriage of quantum theory with Rutherford's nuclear atom was a triumph, explaining a whole host of phenomena, in particular the peculiar fact that atoms of different chemical elements all give off and absorb characteristic wavelengths of light. In time, Bohr's theory would lead inexorably to a revolutionary new description of the subatomic world: quantum mechanics. (Much more on that later.)

Augmented by Bohr, Rutherford's atomic model finally allowed physicists to solve the riddle of the periodic table. In the months following Rutherford's first paper on the nuclear atom, a young research student working under Rutherford in the Manchester lab, Henry Moseley, made another profound discovery. When Mendeleev had created the periodic table, he had given each element a label known as the 'atomic number', which simply recorded the order in which the elements were listed. Hydrogen, the lightest element, sat at position 1, helium the next heaviest at number 2, all the way up to uranium at position 92. This number seemed to be closely connected to the masses of the different elements, as in general the masses of the elements increased as you moved through the table. However, this wasn't always the case. There were a few instances where the chemical properties of the elements had guided Mendeleev to place a heavier element in front of a lighter element. For example, cobalt (atomic number 27) came before nickel (atomic number 28), despite cobalt having a bigger atomic mass. This atomic number was thought to be nothing more than a useful label with no physical significance, but Moseley found that the frequencies of X-rays emitted by different chemical elements depended directly on the atomic number, not the atomic mass. The atomic number was more than just a label after all; it was actually the number of positive charges in the nucleus! So hydrogen contained one positive charge and uranium ninety-two positive charges. These positive charges were balanced by an equal number of negatively charged electrons in orbit around the nucleus, leaving the atom neutral overall. The patterns Mendeleev had first spotted while he was playing cards on those long train journeys through Russia were all to do with the way these different numbers of electrons arranged themselves around the nucleus.

But all this raised a question: what was the nucleus made from? Was Prout's idea that all atoms were made of hydrogen right after all? The fact that the charge of the nucleus was always a whole number multiple of the charge of the hydrogen atom seemed to suggest so, but then, radioactive decay released helium nuclei and electrons, so maybe the nucleus was made of them as well? While much of the physics community embraced the weird and wonderful new world of quantum theory, Rutherford led a relatively small group of physicists on a new journey of exploration. This time their goal was to unlock the make-up of the atomic nucleus itself.

CHAPTER 4

Smashed Nuclei

How are we doing in our quest to make an apple pie from scratch? Well, we've certainly got a better understanding of the basic ingredients. The carbon, oxygen, and hydrogen that we extracted at the outset are made of distinct atoms, and every atom has the same essential structure: an unimaginably tiny nucleus containing almost all of the atom's mass, with much lighter sub-atomic particles called electrons buzzing around it. The whole thing is held together by powerful electrical forces, which bind the nega-tively charged electrons to the positively charged nucleus, while some mysterious quantum wizardry stops the whole thing from collapsing in on itself and taking all matter in the universe with it.

We've also figured out what makes a carbon atom different from, say, an oxygen atom – it's all down to the number of positive charges in the nucleus, which attract an equal number of negative electrons to leave the atom neutral overall. Hydrogen, we now know, is the simplest atom of all – its nucleus has a charge of +1 and is orbited by a single electron. Carbon, on the other hand, has six positive charges in the nucleus and six electrons, while oxygen has eight. Moseley discovered that the number of positive charges in the nucleus turns out to be exactly the same as the atomic number, which chemists used to think was just a label telling you where to find a given element in the periodic table. Since the chemical properties of the elements vary in a regular way as you move through the table, this tells us the chemical properties of an atom must be completely determined by the number of positive charges in the nucleus.

Now that's all bloody marvellous, but despite the discovery of the electron and the nucleus, we still don't know how to make hydrogen, carbon, oxygen, or indeed any of the other elements in an apple pie. If it's the number of positive charges in the nucleus that determines whether an atom is carbon or uranium, we need to figure out what lies inside the atomic nucleus if we're going to find recipes for all the elements in the periodic table.

When the Rutherford–Bohr model of the atom was established around 1913, the nucleus remained shrouded in mystery. However, it was pretty clear that the nucleus wasn't just a new version of the indivisible atom, shrunk tens of thousands of times in size. Marie Curie and Ernest Rutherford were both convinced that alpha, beta, and gamma radiation emerged from the nucleus itself, which meant that the nucleus must be made of even smaller things. The question was, what?

Since the alpha and beta particles that came flying out of the nucleus in radioactive decay were just helium nuclei and electrons, it was natural to assume that the nucleus contained helium nuclei and electrons. Lingering in the background was William Prout's old idea that hydrogen atoms were the basic building blocks of all the heavier elements, but that idea had hit the rocks thanks to awkward elements like chlorine, whose atomic masses weren't whole number multiples of the mass of hydrogen.

It was all a bit of a muddle. Before physicists could make any progress, they were going to need some new experimental clues, but getting information out of the nucleus was going to be no easy task. It would take two more truly heroic experiments, the first of which was carried out by the same scientist who discovered the nucleus, that boisterous, booming force of nature, Ernest Rutherford.

THE NUCLEUS SPLINTERS

In 1914, as the outbreak of war brought scientific research to a grinding halt across Europe, Rutherford abandoned his radioactive experiments to work on submarine detection for the Admiralty. However, even a world war couldn't keep him away from his one true love for long. Although now in his mid-forties and recently elevated to Sir Ernest Rutherford, his curiosity was as strong as ever. The discovery of the nucleus had opened up a brand-new frontier and he was itching to explore it.

His keen scientific nose had already caught the scent: a nagging problem that could be traced back to that Sunday evening in 1910 when he had first shared his vision of the nucleus with the young Charles Galton Darwin. During their after-dinner chat, Darwin had pointed out that Rutherford's idea implied that if you fired alpha particles into gases of light elements like hydrogen, then occasionally the much lighter hydrogen nuclei should be knocked out of the gas, like a snooker ball struck by the cue ball.

Just before the war broke out, Ernest Marsden, the young researcher who had first seen alpha particles bouncing backwards off gold atoms, had followed up on Darwin's suggestion by firing alpha particles into ordinary air. Now air contains a certain amount of water vapour (H_2O), which of course contains hydrogen atoms, and just as Darwin had predicted, Marsden saw hydrogen nuclei being kicked out of the gas by the alpha particles. However, he was puzzled to find many more hydrogen nuclei come flying out than he expected based on the amount of water vapour in air. Somewhat stumped, Marsden eventually made the rather unconvincing suggestion that the radium atoms that produced the alpha particles must also be shooting out hydrogen nuclei.

Rutherford wasn't persuaded. Unfortunately, Marsden had left Manchester for a university post in Wellington, New Zealand, in 1915, only to return to Europe to fight for the British Army in France. After writing to ask for Marsden's permission, Rutherford picked up where his former student had left off, gradually spending more and more time on the experiments as the war ground on. The once bustling Manchester lab was now more or less deserted, and Rutherford found himself working down in a dark basement, accompanied only by the laboratory steward, William Kay.

The apparatus he worked with was an absolute classic of Rutherfordian simplicity: a battered brass box about 10 centimetres long with a radioactive lump of radium at one end and some pipes that could be used to feed in various gases. At the end farthest from the radium was a small window covered in a thin metal foil, which blocked alpha particles emitted by the radium but allowed the more penetrating hydrogen nuclei to escape. Just outside the window was a zinc-sulphide screen, which produced characteristic flashes of light when it was hit by the escaping hydrogen nuclei.

Again, the observations were made in almost total darkness, peering down a microscope at the zinc-sulphide screen. It was eye-straining work: the flickers of light produced by hydrogen were much fainter than those from alpha particles, and an observer could only keep counting for a couple of minutes before their eyesight became unreliable. It was even possible to fool yourself into thinking you'd seen a hydrogen flash if you watched for too long. Rutherford and Kay worked in shifts of around two minutes each, one counting while the other rested his eyes. Rutherford's notebooks from the time tell a story of numerous experimental difficulties, from stray light reflecting off the metal foils to suspected contamination in the gas supplies, including frequent remarks like 'No observations because of poor eyesight.'

For a long time, he struggled to make sense of what he was seeing. Were the hydrogen nuclei coming from contamination in the gas? Perhaps they were somehow being produced when the alpha particles hit the metal foil at the end of the brass box? Or maybe they came from the radium itself as Marsden had suggested? Again, he was forced to put his work on hiatus to go on a mission to the United States in the summer of 1917, but it turned out to be one of those useful breaks when stepping away from a problem lets your mind slowly work out the solution in the background. When Rutherford got back to the lab in September, he had the answer – the hydrogen nuclei weren't already in the gas, they were being *created* when alpha particles collided with atomic nuclei in the gas.

In a burst of intense work in October and November, Rutherford tried out a variety of different gases, from ordinary air to pure carbon dioxide, nitrogen, and oxygen. When alphas were fired into air, the screen lit up with the flickers of hydrogen nuclei, but when he used pure carbon dioxide or oxygen, he saw almost no flashes at all. Pure nitrogen, on the other hand, sent an even larger torrent of hydrogen nuclei careering into the screen than ordinary air. Having eliminated every other possibility, Rutherford was forced to a staggering conclusion: the alpha particles were smashing the nitrogen nuclei apart, sending hydrogen nuclei flying out like shrapnel from an explosion. Well aware of just how momentous this discovery was, he completely absented himself from his submarine work, writing to his superiors at the Admiralty, 'If, as I have reason to believe, I have disintegrated the nucleus of the atom, this is of greater significance than the war.'

After a year of checking and rechecking his results, he felt ready to draw his final, dramatic conclusion: 'The hydrogen atom which is liberated formed a constituent part of the nitrogen nucleus.' Rutherford had finally found the first convincing evidence that the chemical elements were all ultimately made from hydrogen. He would later give the hydrogen nucleus a new name, confirming its status alongside the electron as one of the fundamental building blocks of all atoms, calling it the 'proton'.* It was the climax of a long story that goes all the way back to John Dalton's measurements

---

* The word 'proton' was inspired by William Prout's hypothesis that the chemical elements were all made from hydrogen atoms, which he dubbed 'protyls' when he had first published the idea in 1815.

of the relative masses of different atoms, which in turn had inspired William Prout to imagine that all the chemical elements might be built from hydrogen. Not only had Rutherford resurrected Prout's hypothesis, he had opened a door to understanding the ultimate origins of the chemical elements. Armed with the proton and the electron, physicists could at last begin to imagine how the elements might be built up one by one, from helium all the way to uranium. And he had achieved all this despite wartime shortages, working down in the basement of a deserted laboratory with nothing more than a battered brass box, a few crumbs of radium, and the loyal assistance of William Kay.

But there was a dirty great fly in the ointment: the riddle of elements like chlorine. If all the elements were really made of hydrogen, why did chlorine have an atomic mass 35.5 times that of hydrogen? In fact, a possible way out of this conundrum had already been found by Rutherford's old colleague from his days at McGill, Frederick Soddy. In 1913, Soddy made the curious discovery of a number of new radioactive elements that appeared to be chemically indistinguishable from other well-known *nonradioactive* elements. Among them was a radioactive form of lead, the garden variety of which isn't radioactive at all. That must mean that there could be multiple versions of the same chemical element, occupying the same slot in the periodic table but differing in their radioactivity. Soddy dubbed these chemical copies 'isotopes'.

Now this all raised a tantalizing possibility: what if Soddy's isotopes had the same nuclear charge, making them the same chemical element, but different atomic masses? Perhaps there were really two different isotopes of chlorine, one with a mass of 35 and another with a mass of 36, that when mixed together made it look as though chlorine had an atomic mass in between the two. It was an appealing idea, but it was hard to imagine how you could separate two different isotopes and measure their atomic masses. After all, isotopes were, by definition, chemically indistinguishable.

However, there is a way to do it. Back at the Cavendish

Laboratory, the chemist Francis Aston had been beavering away down in a gloomy cellar on a brand-new instrument capable of weighing atoms with unheralded precision: the mass spectrograph. Aston's new invention could be used to fire ions of different elements through a sort of lens made from electric and magnetic fields, focusing them onto different locations on a photographic strip depending on their masses.

The spectrograph was a revelation, and Aston soon used it to show that chlorine, the element that had scuppered the idea that hydrogen was the building block of all atoms, was actually a mixture of two isotopes: roughly three parts chlorine-35 to one part chlorine-37, giving an average mass of 35.5. By 1922 he had discovered forty-eight different isotopes of twenty-seven different elements, including six different isotopes of xenon alone. All of the elements that Aston weighed turned out to have masses that were whole number multiples of hydrogen's,* a spectacular result that, combined with Rutherford's coup of knocking protons out of the nucleus, all but confirmed that protons were the building blocks of the nucleus.

Between them, Rutherford and Aston had created the first truly unified theory of matter, radical in its simplicity. You only needed two ingredients to make any atom you fancied: the proton and the electron. Protons are just the nuclei of hydrogen atoms and are positively charged, while electrons are negative and have a tiny mass, two thousand times lighter than the proton. At the time, Rutherford, Aston, and most other physicists believed that the tiny nucleus must contain both protons and electrons squashed together inside it, with other 'atomic' electrons orbiting the nucleus at much larger distances. You needed to have electrons in the nucleus to

---

* There was one puzzling exception – hydrogen itself, which came out with a mass of 1.008. That little extra mass turns out to be a source of sunlight, starlight, and ultimately all the other elements in the universe, as we'll see in Chapter 5.

explain why all the elements except hydrogen had masses that were roughly twice as big as their positive charges. For example, the helium nucleus has a charge of +2 but weighs as much as four hydrogen atoms. This must mean that the helium nucleus contains four positive protons combined with two negative electrons, which cancel out those two extra positive charges. To make a helium *atom* you then add two extra atomic electrons in orbit around the nucleus, leaving the atom electrically neutral overall. The same goes for carbon, which was thought to have a nucleus made of twelve protons and six electrons, giving it a mass of 12 and a charge of +6, with six further electrons orbiting the nucleus to complete the carbon atom. What's more, the fact that radioactive elements occasionally spat out electrons (known as 'beta rays') only added to the evidence that electrons must exist inside the nucleus.

As for isotopes, these could now be understood simply by adding extra protons and electrons to the nucleus. Add two extra protons and two extra electrons to the nucleus of chlorine-35 and you've got chlorine-37. The electric charges of the protons and electrons cancel each other out, meaning that the total charge of the chlorine nucleus stays the same (which, after all, is what sets the atom's chemical properties and means it's still chlorine), but we've successfully added two units of mass to the nucleus to create a heavier version of the same atom.

The theory was a triumph, neatly explaining the make-up of the chemical elements, how atoms change in radioactive decay, and why many of the elements have different isotopes. Unfortunately, it was wrong. Rutherford, Aston, and their colleagues were missing a key ingredient, the one we need to finally complete our atomic shopping list and allow us to make any chemical element we like. Finding it, though, was going to be a long and tortuous process.

NEUTRON, WHERE ART THOU?

In a room at the Cavendish Laboratory, two full-grown men sit hunched in what can only be described as a large box. One is Ernest Rutherford, the big-boned, booming father of nuclear physics. Squashed in beside him is James Chadwick, pale, thin, and taciturn. They make an odd couple. Outside, the lab assistant, George Crowe, has just brought down a radioactive source from the storeroom in the Cavendish's neo-Gothic tower and is now busily setting up the apparatus for their experiment. As they sit together in the darkness waiting for their eyes to adjust, naturally they talk.

Since his return to Cambridge to take charge of the Cavendish following J. J. Thomson's retirement, Rutherford has been pondering the same question that we're trying to answer: how do you make the chemical elements? He has realized that as you start to build heavier and heavier atoms by adding protons to the nucleus you soon run into a serious problem. As the nucleus gets bigger, so does its positive electric charge, which means it exerts a stronger and stronger repulsive force on any proton that tries to get close. Eventually this force becomes so enormous that for a proton to get into the nucleus it would have to be travelling at what Rutherford considered to be impossibly high speeds.

Now Rutherford is not usually one for wild speculation, but here, sitting with Chadwick in the dark, he lets his imagination wander. If both electrons and protons exist inside the nucleus, then why shouldn't it be possible to squeeze a single electron and a single proton together to make a nucleus with an electric charge of zero? This neutral nucleus would be unlike any particle seen so far. It wouldn't form atoms in the traditional sense, would be totally chemically inert, and would be impossible to contain in any vessel. But this weird hypothetical particle might just hold the key to the creation of all the elements. We know it today as the neutron.

While a positively charged proton is repelled by a positively charged nucleus, the neutron would face no such obstacle. No electric charge means no repulsive force; the neutron could simply breeze its way into any nucleus, even one with a gigantic repulsive field around it, a bit like a ghost walking through the wall of a heavily fortified castle. As they talk Rutherford and Chadwick become convinced that adding neutrons to the nucleus is the only way to build up the heavier atoms – without the neutron, most of the elements in the periodic table simply would not exist.

But finding the neutron, if it exists, is going to be devilishly hard. Every method of detecting particles that exists at the time relies on the particles' electric charge to make them visible in one way or another. Protons and alpha particles only produce flashes when they hit a zinc-sulphide screen, thanks to their electric charge. The neutron, on the other hand, would leave no trace at all.

The first experiment they try is like something out of *Frankenstein*. Rutherford supposed that if you could pass an extremely powerful electric arc through a tube of hydrogen gas, then perhaps the huge electrical forces would drive electrons and protons together and neutrons would come flying out. They try it, at not inconsiderable risk to their safety, but with no success. In fact, every experiment they try ends in failure.

As the 1920s wore on, Rutherford and Chadwick cooked up ever-more desperate schemes to trap the elusive neutron. As Chadwick later put it, 'I did quite a number of quite silly experiments, when it comes to that. I must say the silliest were done by Rutherford.' For the first time in his life, Rutherford was firing blanks. The indefatigable bloodhound of physics became increasingly frustrated and disillusioned, spending less and less time in the lab and devoting more of his energy to his growing national and international role as a scientific leader. Appointed assistant director of the Cavendish, Chadwick was now responsible for the day-to-day running of the lab, setting projects for the researchers and fighting a continual battle against the lack of equipment and space. By the mid-1920s the Cavendish was beginning to show its age, while

Rutherford stubbornly insisted on filling the creaking building with as many research students as possible.

Although he was undoubtedly an inspiring director, Rutherford's belief that any experiment worth its salt could be done on a shoestring budget was beginning to hamstring the Cavendish's work. After all, who needs fancy apparatus? He had unravelled the mysteries of radioactivity, discovered the atomic nucleus and broken it apart using startlingly basic equipment that could fit easily on a lab bench. When one student complained that he didn't have the equipment necessary to make progress, Rutherford bellowed, 'Why, I could do research at the North Pole!' This attitude was now also starting to put a strain on his relationship with Chadwick.

Chadwick was by no means unresourceful – during the First World War he had managed to run a makeshift laboratory while a prisoner at the famous Ruhleben camp in Germany – but even he struggled to meet the needs of his researchers. He later recalled how one young Australian physicist, Mark Oliphant, once came to him almost in tears, unable to make any progress at all without the right sort of pump. The only way Chadwick could placate the distraught young man was to 'borrow' a pump from Rutherford's personal research room, which he kept strictly reserved for his own public demonstrations.

Nonetheless, Chadwick soldiered on. He was convinced the neutron must be out there; it was only a matter of finding the right experiment. 'I just kept on pegging away,' he later recalled. 'I didn't see any other way of building up nuclei.'

When Rutherford had returned to Cambridge in 1919, most of the global physics community was caught up in the quantum revolution that was shaking physics to its foundations. Studying the atomic nucleus, on the other hand, was a bit of a fringe pursuit. Rutherford had taken the Cavendish out on a limb as the only lab dedicated almost entirely to nuclear physics, but by the end of the 1920s researchers in Vienna, Berlin, and Paris were starting to challenge Rutherford's lab for the crown.

A new way to peer into the nucleus had caught their interest. When nuclei of lighter elements were bombarded with alpha particles, they often emitted high-energy particles of light, known as 'gamma rays'. The idea was that when an alpha particle smacked into a nucleus it briefly kicked its constituent protons and electrons out of their usual positions into an 'excited' state. Almost immediately, these electrons and protons would then fall back into a more stable arrangement, giving off gamma rays in the process. Physicists realized that these gamma rays could act as messengers from deep within the nucleus, potentially carrying precious information about its internal structure. By studying them they hoped to discover a true theory of the nucleus, including an understanding of the mysterious forces that hold it together.

But there was a problem. Radium, which had been physicists' favourite source of alpha particles since Marie Curie had discovered it back in 1898, also gave off large numbers of gamma rays. This made it very hard for an experimenter to know whether a gamma ray had come from a nucleus that had been knocked out of whack by an alpha particle or directly from the radium source itself. What was needed was a different source of alpha particles, one that gave off far fewer gamma rays. Fortunately, Marie Curie had discovered just such an element back in 1898, which she had named 'polonium' after her home country of Poland. Research at the Cavendish had long been hampered by a shortage of the rare element. The lab with the largest source of polonium in the world by far was Marie Curie's own Institut du Radium in Paris.

The great Marie Curie herself was now an international scientific leader, head of the Paris institute and two-time Nobel Prize winner. Increasingly her work took her away from front-line research, but another Curie was ready to step into her shoes: her daughter Irène.

In the autumn of 1931, Irène's interest was piqued by a paper written by two physicists in Berlin, Walther Bothe and Herbert Becker. Bothe and Becker had been bombarding light atoms (all the elements from lithium to oxygen along with magnesium, aluminium, and silver) with alpha particles produced by polonium and studying the gamma rays that came flying out. However, when they got to beryllium,* they had seen something strange: gamma rays that were able to punch through a 7-centimetre-thick iron plate. Normally that amount of iron would have stopped a gamma ray dead. Stranger still, far more gamma rays came flying out from beryllium than from the other elements they'd tested.

Irène had a big advantage over the Berlin team – a polonium source that was ten times more intense. Working with her husband and scientific partner, Frédéric Joliot, she quickly repeated Bothe and Becker's experiments, finding that the gamma rays emitted by beryllium were even more penetrating than her German colleagues had thought. However, most surprisingly, they found that when these gamma rays were fired at paraffin wax, protons came whizzing out at tremendous speeds.

Think of it like a trick shot in a game of nuclear billiards: one ball smacks into another, which collides with another, and so on. We start with radioactive polonium, which fires out alpha particles. These alpha particles smack into beryllium nuclei, which then emit some kind of highly penetrating radiation that Irène and Frédéric assumed to be gamma rays. These gamma rays then fly into a sample of paraffin wax, which is a compound containing lots of hydrogen atoms. The gamma rays knock some of the hydrogen nuclei out of the paraffin, which emerge as high-energy protons.

The most surprising thing about all of this was the incredible energies that the protons got accelerated to when they were struck by one of these gamma rays. To explain the next bit I'm going to have to introduce the idea of something called an 'electron volt'. An electron volt is just a unit of energy like the (possibly) more familiar joule or calorie, but while calories are great if you want to talk about the energy in a slice of apple pie, they're not very

---

* The fourth element in the periodic table after hydrogen, helium, and lithium – a rare soft silvery metal.

convenient for dealing with subatomic particles. Compared to an atom, a calorie is a stupidly huge amount of energy. Using calories to talk about the energy of a subatomic particle would be like giving your bodyweight in solar masses.* So instead we use units more appropriate to the atomic world – the electron volt (eV) – which is just the energy of an electron that's been accelerated using a 1-volt battery.

Curie calculated that to accelerate the protons to the speeds she measured, the gamma rays had to have absolutely enormous energies – around 50 million electron volts (MeV)! This was very difficult to make sense of; the alpha particles emitted by polonium had a maximum energy of around 5.3 MeV. Even if a beryllium nucleus swallowed the alpha particle whole, how on earth could it then release a gamma ray with ten times more energy than it had absorbed? Something very strange indeed was going on.

On a cold January morning a few days after Irène Curie had presented her extraordinary results to the French Academy of Sciences, James Chadwick was leafing through the latest scientific journals in his office at the Cavendish Laboratory. Opening a newly arrived copy of the *Comptes rendus*, he read Curie's paper on beryllium radiation with increasing astonishment. A few minutes later, the young physicist Norman Feather burst into Chadwick's office, just as astounded as he was. At around eleven o'clock Chadwick went to tell Rutherford the news from Paris. As he listened, Rutherford's eyes slowly widened in amazement, until he eventually thundered, 'I don't believe it!' Chadwick had never seen his boss get himself into such a state over a scientific paper. They were both convinced by Curie's results – her experiment was just the kind of elegant and simple set-up that Rutherford so admired – but

* One solar mass (the mass of the Sun) is 2 million trillion trillion kilograms, so I weigh about 0.0000000000000000000000000000039 solar masses. Not a very convenient way of quantifying human bulk, I hope you agree, but I guess it does put any worries about one's weight in perspective.

her explanation of what was going on was another matter entirely. Curie had set out to study gamma rays from beryllium and so had never considered that the radiation she was observing might be something other than gamma rays. Chadwick, on the other hand, who had spent eleven years on a futile search for the neutron, immediately saw the significance of the Paris result. Beryllium wasn't emitting gamma rays at all; it was emitting neutrons.

Chadwick realized that if you assumed the radiation given off by beryllium was made of neutrons instead of gamma rays, then the energy problem disappeared. A gamma ray has no mass, and so to knock a massive proton out of paraffin wax it has to have a very high energy. Imagine firing a ping-pong ball at a bowling ball – that ping-pong ball is going to have to be moving incredibly quickly to get the much heavier bowling ball to budge an inch.

The neutron, on the other hand, should have a similar mass to the proton,* which meant hitting a proton with a neutron was more like hitting a bowling ball with another bowling ball. Chadwick calculated that while gamma rays needed to have energies of 50 MeV, a neutron only needed an energy of 4.5 MeV, less than the 5.3 MeV carried by the alpha particles absorbed by the beryllium nucleus. Suddenly, it all made a lot more sense. But still, he needed proof.

Chadwick knew he was in a race against time. Surely it wouldn't be long before Curie or the Berlin group realized the significance of the beryllium results. Locking himself away in the lab with a polonium source scavenged from a hospital in Baltimore, he worked like a man possessed, allowing himself only around three hours' sleep each night for fear that his competitors might be on the same trail. After a decade of failure and frustration he was damned if he was going to be pipped at the post. A fortnight later he emerged, grey faced with exhaustion but triumphant.

---

* Remember that according to Rutherford a neutron is made of a proton and an electron.

In February, Chadwick attended a meeting of the Kapitza Club – a deliberately informal gathering of physicists organized by the exuberant Russian Pyotr Kapitza in his private rooms at Trinity College. Loosened up by a good dinner and a few glasses of wine, the usually reserved Chadwick gave an uncharacteristically confident presentation, using only a piece of chalk and a blackboard. Fielding frequent interruptions from Kapitza and the rapt gathering, he took his audience from the original clue provided by Curie and Joliot to his final conclusion. After weeks of bombarding paraffin wax and a whole host of other materials, Chadwick had conclusively demolished the idea that the mysterious particles emitted by beryllium were gamma rays. For that to be true, the sacred law of the conservation of energy would have to be broken. Curie and Joliot's results, and all his own observations, pointed unambiguously to a neutral particle with a mass close to that of the proton. The rumours that had been swirling around the Cavendish for the past few weeks were true. After more than a decade of fruitless struggle, Chadwick had discovered the final and most elusive building block of the atom, the neutron.

After such a long drought of new discoveries, Rutherford and the Cavendish Laboratory as a whole basked in the glow of Chadwick's triumph. Shortly after Chadwick submitted an account to the journal *Nature*, Rutherford publicized the discovery during a lecture at the Royal Institution in London, just as his old boss J.J. had done when he first caught a whiff of the electron back in 1897. The discovery of the neutron was particularly sweet for Rutherford – after all, he was the one who had first predicted its existence a dozen years earlier, back in 1920.

However, it was not a total victory. Chadwick had attempted to measure the mass of the neutron, finding that it weighed slightly less than a proton. Rather counter-intuitively, this actually supported Rutherford's idea that a neutron was made of a proton and an electron, as for the neutron to be stable, some energy must be released when the proton and the electron fuse. This 'binding energy' has the effect of actually making the combination weigh less than the sum of its parts.

But back in Paris, Irène and Frédéric hadn't given up their work on beryllium. Using a more precise method, the Paris pair were able to show that Chadwick got the mass of the neutron wrong; it actually weighed about 0.1 per cent *more* than a proton. Rutherford was eventually forced to admit that the neutron wasn't made of a proton and an electron after all.

In fact, the whole idea that the nucleus was made of protons and electrons was wrong too. Physicists had fallen for the logical fallacy that because electrons come out of the nucleus, they must have been inside the nucleus to start with. It turned out that electrons are actually created at the moment a nucleus undergoes radioactive decay. The atomic nucleus is made not of protons and electrons, but of protons and neutrons. During radioactive beta decay, a neutron in the atomic nucleus transforms into a positively charged proton, which stays inside the nucleus and shoots out a negatively charged electron.

The neutron was soon promoted to a fundamental building block of the atom in its own right, alongside the proton and the electron. With these three particles you can make any atom you like, from hydrogen (1 proton and 1 electron) to uranium (92 protons, 92 electrons, and 146 neutrons). The question now is, how do these ingredients actually come together to make the chemical elements in our apple pie? To answer that question, physicists would have to look to the stars.

CHAPTER 5

Thermonuclear Ovens

A couple of years ago I passed through the sleepy English village of Culham on my way to visit one of the largest nuclear experiments in the world. Nestled in a sinuous curve of the upper reaches of the River Thames and surrounded by picturesque Oxfordshire countryside, Culham seems an unlikely place to find scientists struggling to control one of the most powerful forces in the universe. A short drive from the village is a sprawling science campus where an international team are trying to pull off a truly Promethean feat: they are trying to make a star on Earth.

I was met at reception by Chris Warrick, the lab's head of communications, who had kindly agreed to be my tour guide for the day. In principle I was there in my capacity as a curator at London's Science Museum to see if there were any exciting bits of scientific equipment I could scavenge for the collections, but it was also a great excuse to see an experiment I had been itching to visit since I was a teenager – the Joint European Torus, or JET for short.

JET is the world's largest nuclear fusion reactor: a huge metal doughnut that heats hydrogen to temperatures of hundreds of millions of degrees. Under these extreme conditions, hydrogen nuclei fuse together to make helium, releasing tiny blasts of heat and light, replicating the energy source of the Sun and stars. The team at JET are working to tame and control this awesome power.

If they can crack it, nuclear fusion could provide enough clean,* cheap energy to meet all our needs for millions of years.

The dream of harnessing the power of the stars on Earth has inspired scientists and engineers since nuclear energy was first discovered in the 1930s. Today, the climate crisis makes the prize immeasurably greater. When I was thinking about whether to do a PhD back in the late 2000s I seriously considered applying to do research on nuclear fusion, and although in the end the chance to work on the newly minted Large Hadron Collider was just too good to pass up, I had always wanted to see a reactor up close.

From reception, Chris led me across a road and into the large white building with a 1960s space-age look that would make it a good double for Starfleet HQ from *Star Trek*. After wending our way through a maze of corridors and security doors we stepped into the main hall. JET towered above us: a mass of pipes, cables, and machinery dominated by eight hulking iron transformer cores that protrude outwards from the central reactor like huge orange buttresses. Confronted with the sheer bulk of the machine I couldn't help but be left with the impression of some fearsome power being restrained within.

As we walked around the reactor, Chris explained the challenges that his colleagues were grappling with. The extreme temperatures needed to achieve fusion make it impossible to contain the burning hydrogen within any solid vessel. Instead it is kept away from the reactor walls by using a powerful magnetic field that forces the hydrogen into a ring running around the centre of the doughnut-shaped reactor. When JET was built in the early 1980s it was hoped that it would be the first fusion experiment to achieve breakeven – the point at which you get more power out of the fusion reactions

---

* Nuclear fusion reactors would produce no carbon dioxide and unlike nuclear fission reactors, which generate their energy by breaking uranium nuclei apart, no long-lived radioactive waste.

than you put in. Unfortunately, this holy grail was never achieved thanks to a bunch of unforeseen effects that only became apparent once the reactor began operating. Instead, JET is now a test bed for an even bigger reactor currently being built in the south of France known as ITER. This €20 billion megaproject is meant to finally demonstrate the viability of nuclear fusion as a power source, but it has been beset by technical and political problems, leaving some doubting whether it will achieve its goal.

Chris and I sat in his office later that day discussing the prospects for fusion power. He for one remains convinced that we will eventually get there. Slowly but surely, the technical challenges are being overcome and, if nothing else, the promise of limitless clean energy is just too good to give up on. That said, the engineering hurdles remain formidable.

The problem that the scientists and engineers at Culham are trying to solve is the same one we're now confronted with in our search for the ultimate apple pie recipe. Having got our hands on the basic ingredients of all atoms – electrons, protons, and neutrons – we now need to find a way to fuse them together to make the chemical elements in an apple pie. Hydrogen is easy – just take some protons and electrons, shake thoroughly. Carbon and oxygen, which have nuclei made of six protons and six neutrons and eight protons and eight neutrons, respectively, are going to be more of a challenge.

In fact, before we can even begin to think about how to make carbon and oxygen we need to find a way to make an element that apple pies don't contain at all: helium. As the second element in the periodic table, with a nucleus made of two protons and two neutrons, there is no route to carbon and oxygen that doesn't first go through helium.

Unfortunately, as the good people at JET will tell you, making helium from hydrogen turns out to be bloody difficult. To understand why, I'd like to indulge in a little thought experiment. Imagine, if you will, that we are in a nuclear kitchen. In front of us on the work surface are two bowls containing our basic ingredients:

protons and neutrons. Our dish of the day is the helium nucleus, a simple combination of two protons and two neutrons. This is nuclear cooking 101. What could be simpler?

As we saw earlier, the thing that makes a helium atom a helium atom is its nuclear charge – in other words the number of protons in the nucleus – so let's begin by picking up two protons. As we start to bring them together we immediately encounter a problem. The two positively charged particles start to repel each other, and as we force them closer and closer together that repulsive force becomes stronger and stronger. Electrical repulsion between two charges follows what is known as an inverse square law – in other words, the force between two charges quadruples every time you halve the distance between them. This means that long before we get the protons anywhere near touching, a terrific force sends them slipping out of our hands and flying across the room, perhaps smashing some nuclear crockery in the process.

This is precisely the problem that led Ernest Rutherford to speculate about the existence of the neutron. A neutral particle encounters no repulsive force and so bringing a proton and a neutron together should be a doddle by comparison. However, as we turn to the bowl of neutrons we discover to our dismay that while we were looking the other way almost all of them have disappeared, leaving only protons and electrons behind.

This is the second big problem – neutrons are unstable. Outside the safe confines of a nucleus a neutron lives a short and uncertain existence, surviving for only fifteen minutes on average until it spontaneously decays into a proton, an electron, and a third ghostly particle called a 'neutrino' (more on that shortly). Ironically, this instability means that although neutrons were invented to explain how the elements get made, today they play almost no role in forming elements lighter than iron.* They just don't hang around long enough.

---

* We'll see later they do have a role to play in making elements beyond iron, including gold.

We seem to have reached an impasse. The only way forward is to find some way of overcoming the vast electrical repulsion that conspires to keep two protons apart. In fact, we need two different things to happen. First of all, we need another force, an attractive one, that will bind the protons together if we can only get them close enough. The first hints of such a force were found by James Chadwick and a young physicist named Étienne Bieler in 1921. While bouncing alpha particles off hydrogen nuclei they discovered that when they got the particles and nuclei within a few thousandths of a trillionth of a metre of each other an attractive force started drawing them together. This, it turned out, was the first sign of an entirely new force of nature – the strong nuclear force – so called because it is strong enough to overcome the enormous electrical repulsion between two protons.

In the 1920s physicists had almost no understanding of the strong nuclear force, beyond the fact that it must exist to explain how the nucleus holds together and that it only starts to act when two protons are within touching distance of each other. This leads us to the second piece of the puzzle. If we are to fuse protons together to make helium, then we need to find a way to get them close enough for the strong nuclear force to kick into action. However, at this distance – around a thousandth of a trillionth of a metre – the electrical repulsion between two protons is stupendously large: equal to the pull of the Earth's gravity on a 5-kilogram dumbbell. That may not sound like a lot, but bear in mind this is the force on a *single* proton, and a proton's mass is just 0.0000000000000000000000000017 kilograms.

You can think of the repulsive electric field surrounding the nucleus as like the steeply rising ramparts of a heavily fortified castle. To storm the keep, a proton needs to be moving fast enough so that it can 'jump' to the top of the walls, at which point the strong nuclear force takes over and pulls it into the nucleus. This can only happen if the protons are moving fantastically quickly, and such high speeds require terrifically high temperatures,

temperatures of tens of millions of degrees. This is precisely why the scientists at JET need to heat the hydrogen to such enormous temperatures, and why mastering nuclear fusion has proved to be so difficult. However, even if we haven't figured out how to do it on Earth yet, there are places in the universe where such temperatures do exist.

THE IMPOSSIBLE SUN

The first person to come up with a decent estimate of the temperature at the centre of a star was the English astronomer Arthur Stanley Eddington. Eddington's love affair with astronomy had begun in 1886 during night-time walks with his mother along the promenade in the seaside town of Weston-super-Mare. Four-year-old Arthur would gaze up into the inky black and try to count all the stars in the night sky.

By 1920, as director of the Cambridge Observatory, Eddington was grappling with the age-old mystery of the Sun and stars – why do they shine? In total, the Sun blasts 383 trillion trillion watts of power continuously out into space, enough to keep 150 billion trillion kettles permanently on the boil. That's a lot of cups of tea.

Since the mid-nineteenth century a debate had been raging over the source of this tremendous power and, crucially, how long it could keep the Sun shining. On one side were the geologists and naturalists, including the great Charles Darwin, who argued that the Earth and Sun must be hundreds of millions, perhaps even billions, of years old in order to explain the agonizingly slow processes of rock formation and evolution of living things by natural selection. Arrayed against them were the physicists, led by Lord Kelvin, the chap with his own temperature scale, who arrogantly dismissed this as nonsense. There was simply no known power source that could keep the Sun shining at its current rate for more than a few million years, and who were these rock botherers to argue with the laws of physics?

After decades of confusion a vital clue had arrived in 1919. Just down the road from the tranquillity of the tree-lined grounds of Eddington's observatory on the edge of Cambridge, Francis Aston had been labouring away in a dingy basement at the Cavendish Laboratory, weighing atoms using his newly invented mass spectrograph. Aston's great triumph had been to show that every atom had a mass that was equal to a whole number of hydrogen atoms, providing convincing evidence that hydrogen nuclei (protons) were the building blocks of atoms. But there was one puzzling exception to this whole number rule.

Aston's spectrograph could only weigh atoms relative to one another, and so you needed to choose a reference element to compare all your other weights to. In those days, oxygen, which has an atomic mass of 16 (eight protons plus eight neutrons), was the reference element of choice, which meant that the basic unit of atomic mass was defined as one-sixteenth the mass of an oxygen atom. On this scale, one element stuck out: hydrogen itself. By rights a hydrogen atom should have had a mass of exactly 1, but instead it came out ever so slightly higher at 1.008.

When Eddington heard about Aston's peculiar result, he immediately realized its significance. If, as Rutherford and Aston argued, all atoms were made of hydrogen, then that little excess mass might just be the true source of the Sun's power. In 1905, Albert Einstein had argued that mass and energy were interchangeable, an idea expressed by the most famous equation in science: $E = mc^2$.*

The speed of light (c) is a very big number (299,792,458 metres per second to be precise), which means that the speed of light squared is a very, very big number. In other words, this equation tells us every kilogram of mass (m) has the potential to unleash a

---

* $E = mc^2$ doesn't actually appear in Einstein's paper. Instead he uses a *combination* of symbols and words to spell out the same relationship.

cataclysmic blast of energy (E). If four hydrogen atoms could be fused together to make helium then that little bit of extra mass from each hydrogen atom would be converted into energy. Eddington worked out that if hydrogen made up just 7 per cent of the Sun then this process of nuclear fusion could easily keep it shining long enough to keep Darwin and the geologists happy.

Eddington was acutely aware that his ideas were speculative; no one had managed to fuse hydrogen to make helium in the lab. The big question now was whether the centre of the Sun was hot enough to overcome the electrical repulsion between protons and force them together. Fortunately, Eddington had recently created the very tool needed to attack this question – the first realistic theoretical model of the internal workings of a star.

Using his model, Eddington calculated the temperature at the very heart of the Sun – a blistering 40 million degrees Celsius. But despite being far, far hotter than any temperature yet created in a lab, it fell well short of the estimated 10 billion degrees needed to get two protons to fuse. As we saw in our nuclear kitchen, two protons would only be able to overcome their enormous electrical repulsion if they were moving at fantastically high speeds, and even at 40 million degrees they wouldn't be moving anywhere near fast enough.

Undeterred, Eddington was certain that the Sun and stars were fusing helium out of hydrogen, famously retorting, 'We do not argue with the critic who urges that the stars are not hot enough for this process; we tell him to go and find *a hotter place.*' (Perhaps the most genteel way that anyone has ever been told to go to hell.) For Eddington to be right, protons must somehow be breaking the accepted laws of physics. Luckily, breaking the laws of physics was all the rage in the early years of the twentieth century thanks to a revolutionary new theory that was turning the subject on its head.

QUANTUM COOKING

My first encounter with the weird and wonderful world of quantum physics was when my parents gave me a slim paperback for my eleventh birthday called *Mr Tompkins in Wonderland*. The book follows the adventures of the titular hero, 'a little clerk of a big city bank' with a habit of dozing off and dreaming of fantastical worlds inspired by physics. Through Mr Tompkins's adventures we get to explore what the world would be like if everyday objects behaved in a quantum way, leading to, among other things, a very confusing game of snooker and concerns about lions and tigers spontaneously appearing outside their enclosures at the zoo. The author of this charming piece of whimsy was one of the most inventive physicists of the twentieth century, George Gamow. It was his insight that would eventually allow physicists to untangle the paradox of nuclear fusion in the stars.

Georgii Antonovich Gamow was born in 1904, in the Ukrainian city of Odessa on the coast of the Black Sea. Even as a little boy he had a fierce curiosity and a healthy disrespect for authority. Aged ten, he began to doubt his priest's claim that communion bread was transubstantiated into the flesh of Christ, and so one Sunday he sneaked a crumb home in his cheek and examined it using a small microscope bought for him by his father. He concluded that the body of Christ had more in common with ordinary bread than it did with human flesh, using a piece of skin sliced off the tip of his own finger for comparison. He later wrote, 'I think this was the experiment which made me a scientist.'

Despite turmoil caused by the First World War and the subsequent Bolshevik Revolution, Gamow managed to get an excellent education, first at Odessa and then at Petrograd, the top university for theoretical physics in the Soviet Union. His big break came in 1928 when he got to spend the summer in Göttingen, Germany, working at the Institute for Theoretical Physics, which was headed by one of the leaders of the quantum revolution, Max Born.

Gamow found the place 'buzzing with excitement', its seminar rooms and cafes crowded with physicists arguing over the consequences of the new theory. However, Gamow liked to work in less crowded fields and so took himself off to the library in search of a problem that he could claim for himself. It was there that he came across a paper by Ernest Rutherford describing experiments where he had fired alpha particles (helium nuclei made of two protons and two neutrons) at uranium. The article left Gamow scratching his head. Rutherford had found it impossible to get alpha particles to penetrate the uranium nucleus, and yet it was well known that uranium spontaneously spat out alpha particles. How could it be that alpha particles couldn't get into the nucleus from outside, and yet ones with only half as much energy could escape from within?

Gamow suspected that the explanation could be found by applying the groundbreaking theory of quantum mechanics to the nucleus, something nobody had tried before. At the time, quantum laws had only been used to explain the way electrons orbit around atoms – it wasn't at all clear if the same rules applied in the mysterious nuclear realm.

At the heart of quantum mechanics is one of the most counterintuitive and yet profound ideas in all of physics: wave–particle duality. At the end of the nineteenth century physicists had believed that light was a wave, a spread-out wibbly-wobbly thing, like a ripple on the surface of a lake. Experiments had conclusively demonstrated light's wavelike behaviour, including its ability to spread out into circular ripples when shone through a small hole, a behaviour known as diffraction, and its ability to *interfere* with itself (not like that) when two light waves combine to make a larger wave or cancel each other out if the crest of one wave meets the trough in another.

However, at the start of the twentieth century things started to get confusing. First off, the German physicist Max Planck showed that it was only possible to explain the colours of light given off by a hot body – say a red-hot piece of iron – if you did your calculations assuming that light came in discrete little packets known as 'quanta'. At first Planck regarded this as a mere mathematical trick to get the right answer, but then, in 1905, Einstein published a paper showing that a puzzling phenomenon known as the 'photoelectric effect' could be explained if light really did come in quantized little lumps. In other words, light was a flow of particles, known as photons.

These two seemingly contradictory assertions about the nature of light kick-started the quantum revolution. At first it was thought that only photons exhibited this weird wave–particle duality, but in 1924 the French physicist Louis de Broglie argued that it wasn't just a property of light; it applied to particles of matter as well. Electrons, protons, and even atoms – particles that had previously been thought of as little hard nuggets with well-defined positions – could be made to behave like spread-out waves too. The year before Gamow arrived in Göttingen, de Broglie's bizarre hypothesis had been dramatically vindicated when George Paget Thomson, son of J.J., fired electrons through thin metal films and discovered that they formed a diffraction pattern,* apparently contradicting his own dad's experiments showing that the electron is a particle.

If all this leaves your head spinning, don't worry. It confused the hell out of the entire physics community well into the 1920s. The most intuitive, or perhaps I should say least counter-intuitive, description of this quantum weirdness was devised by the German theorist Erwin Schrödinger. It's known as 'wave mechanics'.

In general, particles, including photons, electrons, and protons, are detected at specific points in space. For example, if you fire an electron through some experiment at a detection screen, the electron will arrive at a well-defined location on that screen. The fact that the electron appears to arrive in only one location, rather than being

---

* A similar effect was discovered around the same time by two American physicists, Clinton Davisson and Lester Germer at Bell Labs in New York. Davisson shared the Nobel Prize with G. P. Thomson.

spread out all over the place, is what we refer to as the particle bit of its behaviour. However, wave mechanics says that in between the electron being emitted and detected, it doesn't behave like a particle at all – instead, it travels as a wave.

This wave isn't a wave in water, or air, or indeed any other medium, it's a wave of *probability*, known as the 'wavefunction'. The size of the wavefunction is related to the probability of finding the electron at a certain point on the screen; the bigger the wavefunction is at a given point, the more likely it is that we'll find the particle there. Then comes the really mysterious bit. Somehow, the wavefunction *collapses* from being spread out through space, down to *a single point* where the electron is detected. It is impossible to know which point the wave will collapse to in advance; we can only calculate the probability of the wavefunction collapsing at different locations on the screen. This nonsensical process is known as 'wavefunction collapse', and to this day, no one really knows how it works.* All we do know is that this really is how things appear to behave in the subatomic world.

Okay, back to our boy George. Gamow realized that if he used wave mechanics to describe the behaviour of alpha particles shot out by a uranium nucleus, then the paradox that they didn't have enough energy to escape could be overcome. Let's return to our castle analogy, which I actually stole from Gamow's own description in *Mr Tompkins*. In the analogy, the nucleus is like the inside of a castle, protected by a high wall that keeps intruders out and the inhabitants in. Gamow imagines the alpha particle before it escapes the nucleus as bouncing about back and forth inside the castle walls. If we think of the alpha particle in the old-fashioned way, as a hard little sphere, then it doesn't have enough energy to get over the top of the walls and escape.

However, if instead we think of the alpha particle as a wave,

---

* Or indeed agrees if it actually is necessary. For a great account, see *Beyond Weird* by Philip Ball.

then something very strange indeed can happen; it can leak out through the walls, like water seeping through cracks in the brickwork. That leaves a little bit of it outside the castle so that there is now a small but non-negligible probability of the alpha particle being found outside the castle walls. When the wavefunction collapses, the alpha particle can suddenly appear outside the uranium nucleus, as if it has tunnelled through the barrier. It's a bit like a prisoner furiously hurling themselves at their cell wall over and over again until quite suddenly, as if by magic, they pass straight through and find themselves outside and free. Amazingly, there is a tiny, tiny probability that such a thing could really happen to a real prisoner in an actual prison, but the probability of all the atoms in a prisoner's body simultaneously tunnelling through the walls of their cell are so vanishingly small that it will almost certainly never happen, even though in principle it could.

Gamow's theory turned out to be a triumph, unravelling the paradox of how an alpha particle escapes from the uranium nucleus.* While working on his theory over that summer in Göttingen, Gamow struck up a friendship with another physicist one year his junior, the German-born Fritz Houtermans. Gamow and Houtermans clicked instantly; they were both young, charming, and enjoyed reckless bohemian lifestyles, shared a wicked sense of humour that frequently got them into trouble, and were both passionate about physics. Houtermans was very taken with Gamow's alpha decay theory; when he returned to Berlin, he continued to turn it over in his mind.

A few months later, Gamow received a letter from his friend. Back in Berlin, Houtermans had run into a visiting British astrophysicist, Robert Atkinson. While discussing Gamow's theory they had realized that if particles could tunnel out of a nucleus, they should also be able to tunnel *in*. Atkinson was familiar with

---

* The Americans Ronald Gurney and Edward Condon came up with the same solution as Gamow at the same time.

Eddington's work on the temperature at the centre of the Sun and stars and wondered whether nuclear fusion might be possible after all. If protons at the centre of the Sun were able to quantum tunnel through the repulsive electrical barrier that kept them apart, then perhaps nuclear fusion could go on at lower temperatures than previously thought. Maybe, just maybe, Eddington had been right.

The three men settled on the picturesque skiing resort of Zürs in the Austrian Alps as the most agreeable place to work through the theory. Gamow found to his satisfaction that Fritz and Robert 'were almost ready with their calculations, so the discussion did not impose on our skiing time'.

Houtermans and Atkinson's theory was the exact reverse of Gamow's. Instead of a particle inside the nucleus tunnelling out, this time they considered protons hurling themselves against the nucleus's electrical barrier from the outside, like soldiers assaulting the walls of a castle. Eddington's calculations had shown that protons in the Sun weren't moving fast enough to get to the very top of the castle's walls. However, the repulsive barrier around the nucleus gets thinner and thinner the higher up the walls you get. If protons in the centre of the Sun were moving fast enough to climb to a point where the barrier became thin enough, then quantum tunnelling might allow some small fraction of them to jump through and appear inside the nucleus without having to get right over the top.

The question was whether the tunnelling probability was high enough to allow nuclear fusion to go on in the centre of the Sun. After a few days of skiing and drinking, and presumably also a bit of physics, the three arrived at an equation describing the rate of fusion as a function of the temperature and density at the centre of a star. Unfortunately, due to scant knowledge of the make-up of the nucleus in 1929, Gamow got his calculations wrong by a factor of ten thousand. But in one of the most remarkable pieces of dumb luck in the history of science, Houtermans and Atkinson made a second mistake that shifted the answer by a factor of ten thousand in the *opposite direction*. Miraculously, the two mistakes cancelled each other out and the equation they ended up with was essentially correct.

Plugging in Eddington's estimate of the conditions in the solar core, they found to their delight that nuclear fusion really did seem to be possible, and what's more it could easily keep the Sun shining for billions of years.

Houtermans later recounted the climax of their work in typically colourful style. After putting the finishing touches to the article, he went for an evening stroll with Charlotte Riefenstahl, a young physicist whom both he and Robert Oppenheimer* were courting at the time. 'As soon as it grew dark the stars came out, one after another, in all their splendour. "Don't they shine beautifully?" cried my companion. But I simply stuck out my chest and said proudly: "I've known since yesterday why it is that they shine."'

That has got to be one of the best chat-up lines in history. It seems to have worked; Charlotte and Fritz went on to marry, not once, but twice. Houtermans and Atkinson submitted their article under the playful title of 'How to Cook Helium in a Potential Pot'. Unfortunately, a rather unimaginative journal editor renamed it to the decidedly less punchy 'On the Question of the Possibility of the Synthesis of Elements in Stars'.

Titles notwithstanding, their article made little impact, at least at first. Nuclear physics was mired in uncertainty, and what today seem like bonkers ideas were making the rounds. The great Niels Bohr had proposed that the sacred law of the conservation of energy might be broken inside the nucleus, ultimately accounting for the power output of the Sun. To get any traction, Atkinson and Houtermans would need experimental evidence for their tunnelling theory. Fortunately, it would soon be provided by Ernest Rutherford and his team of physicists at the Cavendish Laboratory.

In 1932, Cavendish physicists John Cockcroft and Ernest Walton

---

* The future 'father of the atom bomb'.

used one of the first-ever particle accelerators to bombard a lithium target with a beam of protons, splitting the lithium nucleus in two in the process. This incredible feat was only possible thanks to Gamow's nuclear tunnelling theory. Although Cockcroft and Walton's machine could accelerate protons to an impressive 800,000 volts, that fell well short of the several million volts required to get the protons moving fast enough to get directly over the top of the electrical barrier protecting the lithium nucleus. The only way to explain the fact that they had managed to split the atom was if protons were quantum tunnelling through the barrier, just as Gamow's theory predicted.

With experimental confirmation that quantum mechanics really does apply to the nucleus, the way was finally open to explain how helium gets made inside the Sun and stars. However, there were still some serious obstacles to overcome. We are missing two vital ingredients. One is a rare isotope of hydrogen, the other is beloved of science fiction writers everywhere: antimatter.

HELIUM MADE TWO WAYS

Thanks to the idea of quantum tunnelling we now know that the Sun and stars are hot enough to force two protons together. In other words, we have found the thermonuclear ovens needed to cook helium. But there's a problem. If we're really starting from scratch, then our recipe for helium must surely start by fusing two protons, and there we immediately run into trouble. There is no stable nucleus made of two protons. If there were, it would technically be known as helium-2, but there ain't no such thing.

There is, however, a nucleus made of one proton and one neutron, a heavy isotope of hydrogen known as deuterium, which was discovered by the American chemist Harold Urey in 1931. This gives us a glimmer of hope. What if there were a way of combining two protons and at the same instant transforming one of the protons into a neutron? If we could do that, then we'd be able to make deuterium, the first vital step in our recipe for helium.

Until 1932, turning a proton into a neutron seemed to be impossible. For one thing, where would the positive charge of the proton go? It can't simply vanish out of existence. We're missing a second ingredient, one that was discovered in 1932: the positron. Also known as the antielectron, this new particle is exactly like an electron, except it has a positive electric charge. The positron was the first particle of antimatter to be detected, an incredibly profound discovery that we'll discuss later on, but for now it plays only a small if vital supporting role in our thermonuclear cooking story.

In 1934, physics' Parisian power couple Irène and Frédéric Joliot-Curie discovered a brand-new type of radioactive decay, one in which an unstable nucleus shoots out one of these positrons. They soon realized that deep inside the decaying nucleus a proton had transformed into a neutron. One of the reasons it took so long to discover is that isolated protons can't decay in this way; a proton actually weighs *less* than the neutron it turns into. However, in certain unstable nuclei, a proton can absorb some energy from its host nucleus, allowing it to turn into a heavier neutron, releasing a positron and a neutrino.

Armed with deuterium and a way to turn protons into neutrons, we can at last make progress in our attempt to forge helium. In 1936 Robert Atkinson pointed out a potential first step on the route to building the heavy elements out of hydrogen. At the extreme temperatures present at the centre of the Sun two protons could be forced together, forming for a fantastically brief instant an unstable two-proton nucleus. Then, before it could fall apart, one of the protons converts into a neutron, creating a nucleus of deuterium.

Atkinson's suggestion was the beginning of a period of rapid progress that soon came to a dramatic climax. After several years wandering across Europe, often roaring into quiet university towns on his motorbike, Gamow had defected from the Soviet Union in 1933. Now based at George Washington University in the United

States, he had become increasingly interested in what was known as the 'stellar energy problem' – in other words, why the stars shine – so in 1938 he organized a conference on the topic, inviting thirty-four of the world's top astro-, nuclear, and quantum physicists.

Among them was Hans Bethe, one of the brightest theorists of his generation, who, as Gamow put it, 'knew nothing about the interior of stars, but everything about the interior of the nucleus'. In fact, shortly before the conference, Bethe had been contacted by a former student of Gamow's, Charles Critchfield, who had taken Atkinson's proposed reaction where two protons fuse to make deuterium and run with it, coming up with an entire scheme that started with just protons and through various steps built up a freshly baked helium nucleus. Along the way, Critchfield had run into some mathematical difficulties and so turned to Bethe for help.

Bethe was impressed with the young physicist's work, and after some tweaks to make the calculations more elegant the two of them produced a complete recipe for cooking helium. Today it's known as the 'proton–proton chain'. The modern version of it goes like this:

A RECIPE FOR HELIUM –
THE PROTON–PROTON CHAIN

Step 1: Two protons collide, briefly forming a highly unstable two-proton nucleus.

Step 2: Before the two-proton nucleus can disintegrate, one of the protons decays into a neutron, forming a deuterium nucleus (one proton, one neutron) and releasing a positron and a neutrino.

Step 3: Another proton collides with the newly formed deuterium nucleus to form helium-3 (two protons, one neutron) and releasing a gamma ray.

Step 4: Two helium-3 nuclei smack into each other and
form a nucleus of helium-4 (two protons, two
neutrons), sending the two leftover protons
flying out.

At last, we have a recipe for helium! Even better, the whole process
results in a net release of energy, and so it's also a recipe for starlight. However, there is a problem. Eddington had estimated that
the temperature at the centre of the Sun was 40 million degrees
Celsius, but at that temperature the proton–proton chain would
run too fast, resulting in a Sun that shines far more brightly than
it actually does. Critchfield and Bethe had come agonizingly close
to solving one of the oldest mysteries in science, only to find that
the solar oven was too hot for their recipe.

That all changed at the Washington conference. Bethe's interest
was piqued during a long and detailed talk about the conditions
inside the Sun. When Eddington had come up with his 40-million-
degree figure, people had thought that the Sun was made of more
or less the same stuff as the Earth. However, in 1925 the brilliant
young astronomer Cecilia Payne had shown that the Sun and stars
were dominated by hydrogen and helium, with only relatively tiny
amounts of the heavier elements. When Eddington's calculations
were revised assuming that the Sun was 73 per cent hydrogen and
25 per cent helium, the temperature in the core plummeted to a far
cooler (though still decidedly toasty) 19 million degrees. With the
solar oven set at this lower temperature, Bethe found that the
proton–proton chain predicted a power output for the Sun that was
much closer to the true value.

At long last, the riddle of why the Sun shines had been solved.
Deep inside its core, the crushing force of its own gravity heats
hydrogen to a temperature of 15 million degrees Celsius.* In this
fearsome heat, protons and electrons pinball into one another at

* The accepted modern value.

terrific speeds, and every so often, in one among unnumbered collisions, two protons come close enough for quantum mechanics to take over, creating a tunnel through the repulsive electrical barrier keeping them apart and allowing them to unite to form a nucleus of deuterium. From here, slowly but surely, the Sun builds helium from hydrogen, gradually transforming its own bulk over billions of years, all the while releasing a steady flow of heat that eventually breaks out from the Sun's tortured surface and escapes into space as sunlight. The Sun is a vast thermonuclear furnace.

Our search for the recipe for helium isn't quite finished, though. During the conference, Bethe realized that something was amiss. The proton–proton chain worked brilliantly for stars smaller than the Sun, but when he applied it to larger stars the reaction didn't fit.

Take Sirius, the brightest star in the night sky, a brilliant blue-white jewel in the constellation Canis Major. Sirius's apparent brightness is a result of two factors: first off, it's only a short bus ride away in galactic terms, at just a smidge over 8.6 light-years from the Earth. Second, Sirius is about twice as massive as the Sun, which means the crushing force of gravity heats its core to a higher temperature. This higher temperature means that protons are whizzing about faster, which means they can more easily overcome the repulsive electrical forces keeping them apart, boosting the rate of nuclear fusion.

However, the weird thing is that despite only being twice as massive as the Sun, Sirius is twenty-five times brighter, which couldn't possibly be explained using the proton–proton chain. There must be something else happening in Sirius's core that makes it shine so ferociously.

Bethe started thinking about a completely different sort of reaction. Instead of protons joining together directly to form helium, what if instead they were gobbled up by an existing heavy nucleus, which slowly digested the four protons before spitting them out as a fully formed helium nucleus at the end? The question was, did a heavy nucleus with the right properties to act as a proton digestor exist?

Starting with helium Bethe worked his way along the first row of the periodic table, considering and then discarding elements one by one. Helium itself was no use at all as there was no element with mass 5 and so no way to get further by just adding a proton. Lithium, beryllium, and boron were all too scarce and would be burned up by the reaction too quickly to keep a star shining for very long. Then he came to element six, carbon. It seemed to have exactly the properties he was after. Bethe caught the train back to Cornell with the outline of a solution already in his mind.

Just a couple of weeks later, he had come up with a second recipe for helium. It's known as the 'carbon–nitrogen–oxygen (CNO) cycle', and it goes like this:

A RECIPE FOR HELIUM –
THE CARBON–NITROGEN–OXYGEN CYCLE

Step 1: A proton tunnels into a carbon-12 nucleus creating a new nucleus of nitrogen-13, which then decays into carbon-13, emitting a positron and a neutrino.

Step 2: A second proton tunnels into the carbon-13 nucleus creating nitrogen-14.

Step 3: A third proton tunnels into the nitrogen-14 nucleus to create oxygen-15, which then decays into nitrogen-15, emitting a positron and a neutrino.

Step 4: Finally, a fourth proton tunnels into the nitrogen-15 nucleus, breaking it apart to form a helium-4 nucleus and the same carbon-12 nucleus we started with.

Bethe's reaction was almost miraculous. Through a series of successive collisions, a carbon-12 nucleus was able to effectively swallow up protons and convert them into helium. And best of all, at the end of it you got back the original carbon-12 nucleus, allowing the whole process to start again.

Now, as a carbon-12 nucleus contains six positively charged protons, its repulsive barrier is six times higher than hydrogen's. As a result, for a proton to have a chance of tunnelling into a carbon nucleus it has to be moving at a hell of a lick, making the reaction incredibly sensitive to temperature. In fact, if you double the temperature at the centre of a star, the CNO cycle blasts out 65,000 times more power, explaining why Sirius shines twenty-five times more brilliantly than the Sun despite only being twice as big and a little bit hotter. The CNO cycle is now thought to be the dominant source of starlight for all stars more than 1.2 times as heavy as the Sun.*

So there you have it – at last we have the recipes needed to make helium inside stars. But there is one big challenge left to face: how do we actually *know* this is what is going on inside the Sun?

Until relatively recently, our understanding of how the Sun fuses hydrogen to make helium was based on two different bits of science. From the 1930s onwards, physicists started to use particle accelerators to fire protons at various targets, recreating the nuclear fusion reactions dreamed up by Hans Bethe and his colleagues. These pioneering experiments gave physicists a direct handle on how fast each helium cooking process should run depending on the temperature of the stellar oven. Meanwhile, astrophysicists produced ever-more accurate theoretical models that could make increasingly precise estimates of the core temperatures of stars. With these two pieces of crucial scientific knowledge, physicists could infer that stars the size of the Sun were powered mostly by the proton–proton chain, while larger stars like Sirius relied on the CNO cycle.

---

* At the time, Bethe incorrectly believed that the CNO cycle was the dominant power source of the Sun thanks to an overestimate of its core temperature.

However, all the evidence was really only indirect. To know for sure, we need to look straight into a star's burning core and witness the nuclear reactions first-hand. But looking inside a star is impossible, right? When we look at the Sun (with the appropriate equipment, of course – not directly, please) all we can see is its shining surface. The core is hidden, forever beyond our reach.

Or so it seems at first. In fact, it is only in the last few decades that physicists have finally been able to peel back the outer layers of the Sun and stare straight into its heart. Deep inside an Italian mountain a couple of hours' drive from Rome, a team of physicists have built a gigantic detector that patiently watches for ghostly messengers that come to us direct from the Sun's thermonuclear furnace. Their goal is to prove once and for all that the nuclear reactions first proposed in the late 1930s really are the ultimate source of the Sun's awesome power.

SUNSHINE UNDER A MOUNTAIN

On a stiflingly hot August day, I turned off Autostrada A24 near the Italian village of Assergi and onto an unmarked road running across the lower slopes of the towering Gran Sasso mountains. Whether it was the lack of road markings or the fact that I'd got up at three that morning to catch the early flight to Rome, I had briefly relapsed into driving on the left, only realizing my mistake when another car rounded the corner ahead of me. After a panicked swerve and an apologetic wave to the startled-looking driver, I turned the corner to find the road full of Italian police.

The officers were gathered outside the gates of LNGS, the Laboratori Nazionali del Gran Sasso, the largest underground research facility in the world. Hoping that they hadn't witnessed my eccentric driving I cruised gingerly past the assembled cops, who to my relief made no moves to arrest me. Nonetheless, I was more than a little concerned – had something happened underground? I had read that the lab had been involved in some recent legal trouble that was threatening to shut down some of its experiments but hadn't realized for a second that it was this serious. After parking my hire car out of sight just around the next bend I presented myself at security and asked for Aldo Ianni, the physicist who had agreed to be my guide inside the mountain, hoping that the whole thing wasn't about to be called off.

I was there to visit what has got to be the most extraordinary solar observatory in the world. For one, a cavern a kilometre and a half under a mountain isn't an obvious place to study the Sun. But this is no ordinary observatory. The instrument I was here to see doesn't look at the Sun in light, or even radio waves, but in neutrinos.

Neutrinos are the most elusive of all the fundamental particles. They have almost no mass and no electric charge. This makes detecting them fiendishly difficult. Most particle detectors rely on the fact that charged particles interact with the material of the detector through the electromagnetic force, creating a telltale flash of light or an electric current. However, neutral particles don't interact electromagnetically and so are much harder to spot. For exactly the same reason, James Chadwick had had to struggle through a decade of frustration and failure before he finally cornered the neutron. But even though the neutron has no electric charge, it does at least feel the strong nuclear force, which makes it more likely to collide with other atomic nuclei and make its presence known. Neutrinos, on the other hand, don't even feel the strong nuclear force. The only way they can interact directly with ordinary matter is through the third force that governs the quantum realm, the so-called weak force. As its name suggests, the weak force is, well, weak, which means that the chances of a neutrino bumping into an atom are vanishingly small.

However, while this makes detecting neutrinos extremely challenging, it also makes them the perfect tool to probe the inner workings of the Sun. Deep within its core, nuclear fusion reactions are constantly generating a vast flood of photons (particles of light) as well as neutrinos. Unfortunately for solar physicists, those photons endlessly collide with the superheated gas of protons and electrons that make up the body of the Sun, taking tens of thousands of years to pinball their way to the surface, by which time all the information they originally carried about the nuclear reactions that created them has been lost. Neutrinos, on the other hand, face no such obstacle. To them the Sun's enormous bulk is almost completely invisible; they escape to the surface travelling at the speed of light in a little over two seconds, reaching the Earth around eight minutes and twenty seconds later.

By the time you have finished reading this sentence, around two thousand trillion of these neutrinos will have passed straight through you. Fortunately, we are blissfully unaware of this constant barrage as the weakness of the weak nuclear force ensures hardly any will ever so much as glance off an atom of your body. Still, each of these neutrinos carries precious information about the nuclear reactions going on at the centre of the Sun, if only they could be captured.

The experiment I had come to Italy to see does just that. It's called Borexino, a gigantic tank containing a liquid hydrocarbon housed in a cavern deep within the Gran Sasso mountains. The principle of the experiment is easy to understand, even if it is unbelievably hard to pull off in practice. Among the uncountable trillions of neutrinos that constantly pass through the tank, a tiny fraction will collide with electrons, giving them a kick as they pass by. As an electron recoils from the invisible blow, it excites the surrounding liquid, creating a tiny flicker of light, which is captured by an array of detectors encircling the tank. By counting the number of neutrinos and measuring their energies, the physicists at Borexino are able to watch the Sun fusing hydrogen into helium in real time.

After I had spent a few minutes waiting by the security hut in the midday sunshine, Aldo pulled up in his car and greeted me with a handshake. He explained that the heavy police presence was due to an impromptu visit by the Italian finance minister, and all systems were go for our tour of the experiment. To reach Borexino, we first had to drive back onto Autostrada A24 and then straight into the mountainside through the 10-kilometre motorway tunnel. As we drove, Aldo explained that the Gran Sasso laboratory had first been proposed back in the 1970s while the tunnel was being built, with the three huge experimental halls completed in 1987. Meanwhile I did my best to explain to a slightly bemused-looking Aldo what neutrino physics had to do with apple pie; it turns out Carl Sagan isn't a household name in Italy.

With the lofty peaks of Gran Sasso towering above us, we passed from the bright Italian afternoon sunshine into the darkness of the mountain. Above us was a gigantic mass of dolomite rock more than a kilometre thick, without which the Borexino experiment would be totally impossible. The Earth is under constant bombardment from high-energy cosmic rays from deep space. When they strike the upper atmosphere, they produce a shower of electrically charged particles, many of which get all the way to ground level. This cosmic avalanche would completely swamp the rare neutrino interactions studied by Borexino were it not for the mighty shield of the Gran Sasso mountain range, which absorbs them almost entirely while allowing neutrinos from the Sun to pass straight through.

After several minutes driving through the long autostrada tunnel, we took a turn off into a smaller passage that you could easily miss if you didn't know it was there. In front of us was the entrance to the underground lab, a large stainless-steel door that slowly slid open after Aldo buzzed the intercom. The whole thing felt like entering a Bond villain's mountain lair.

We parked in a side tunnel. Stepping out of the car, I was hit by the coolness of the air and that particular damp, mineral aroma that I've only ever smelt in deep caves. Water dripped from the moss-covered walls of the tunnel as we walked the short distance to present ourselves at security. After signing in and passing me a rather fetching blue hard hat, Aldo led me along another long curving tunnel before we stepped through a steel door into a soaring cavern. We had entered Hall C, the home of Borexino, a huge barrel-vaulted concrete chamber 20 metres across, 18 metres high, and 100 metres long. The low hum of machinery was punctuated by a rhythmic high-pitched chirrup, like the mating call of some giant mechanical cricket. This, Aldo reassured me, was just the sound of the vacuum pumps.

In front of us were two huge cylindrical tanks, each several storeys high, part of the complex plumbing system that feeds Borexino. As we walked towards the towering mass of machinery, Aldo explained that the key challenge that he and his colleagues face comes from natural background radiation. The ground we walk on, the objects that surround us, and even the air we breathe all contain tiny traces of radioactive elements, from uranium and radon to carbon-14. These substances emit a constant background of alpha particles, electrons, and gamma rays. Such low levels are harmless to us but would be fatal to an experiment like Borexino.

Despite its huge size, Borexino only sees a few dozen neutrinos each day thanks to the weakness of their interaction with ordinary matter. Such a tiny signal would be totally overwhelmed by normal levels of background radiation, and so Aldo and his colleagues wage a constant war against radioactive impurities in the system. The job of the huge network of tanks and pipes that we stood beneath is to endlessly clean and purify the various liquids inside the Borexino tank, which are distilled and then cleansed of radioactive contaminants using bubbles of highly purified nitrogen gas, before at last being allowed into the experiment. More than that, the material of every component in Borexino had to be carefully selected, manufactured, and tested to produce as little radioactivity as possible. The result of this enormous effort is one of the lowest levels of radiation ever achieved on Earth.

Climbing a steel gantry, we reached one of Borexino's control rooms, where Aldo stopped to talk with a colleague who was busy working on the experiment. I had no idea what they were saying – my Italian is barely good enough to order a coffee – but his colleague seemed agitated. Aldo explained afterwards that the system used to cool the electronics that read out the data had broken down, and they were working to bring it back online as quickly as possible. Every day of data collecting is precious when you are working with such rare events, and the Borexino team are in a race against time.

Just a few months earlier, at the end of 2018, the Borexino collaboration had published a comprehensive study of the neutrinos produced by the proton–proton chain, the reaction that provides 99 per cent of the Sun's power. As the proton–proton chain slowly builds helium from hydrogen, neutrinos are released whose energies reveal the stage in the process that they came from. After almost two decades of measuring the numbers and energies of the arriving neutrinos in painstaking detail, the scientists at Borexino had found that the fusion reactions first proposed by Hans Bethe and Charles Critchfield way back in 1938 were going on deep in the heart of the Sun just as they had predicted.

However, one piece of the puzzle remains – the CNO cycle, the fusion reaction where carbon gradually swallows protons before spitting out a fully formed helium nucleus. This second reaction only produces 1 per cent of the Sun's power, which makes it much harder to see. The prize, though, is huge. If Aldo and his colleagues can detect neutrinos from the CNO cycle, it would be the final verification of one of the oldest mysteries in science: namely, why the Sun shines. More than that, since the CNO cycle is thought to be the main power supply of all stars 1.2 times heavier than the Sun, seeing it going on in nature in real time would be a spectacular coup.

Unfortunately, at the start of 2019 Borexino ran into an unexpected and potentially terminal problem. Not a scientific one, mind, but a legal one, whose history can be traced all the way back to 2002. That summer, human error led to some of the liquid hydrocarbon used in the detector escaping into the groundwater. Since then, environmental safety standards and procedures have been significantly tightened at the lab and everyone I spoke to was convinced that the risks of a similar incident were now absolutely tiny. But the damage to local community relations was done. A determined, decade-long campaign by environmental activists finally came to a head just two months before my visit, when it was announced that three senior managers at LNGS would face criminal prosecution. Another consequence of the court case is that all new experimental work has been suspended and Borexino will have to shut down in a little under two years.

Hence the race against time. The big question in the researchers' minds is whether two years will be enough to see neutrinos from the CNO cycle. The rarity of CNO neutrinos means that to have any chance of spotting them the team at Borexino will have to control radioactive backgrounds at an unprecedented level.

Walking farther along the gantry several metres above the cavern floor we passed the control room and moved towards the back of the experimental hall. The mechanical chirruping grew louder until we finally came face-to-face with Borexino itself: a 17-metre-tall domed tank, covered in silver insulating foil that glistened softly in the artificial light. Towards the top, loops of blue piping were wrapped around its girthy 18-metre diameter. The whole thing looked like some nineteenth-century vision of an alien spacecraft. The shiny insulation and blue piping, Aldo told me, had only been added recently in the hope that it might make seeing CNO neutrinos possible at long last.

When Borexino was first proposed way back in the 1990s, no one had thought that it had any chance of seeing neutrinos from the CNO cycle. The signal was just too weak and the radioactive backgrounds too high. However, in recent years the team realized that a peculiar feature of the underground environment might make it possible after all.

The surrounding mountain rock keeps the cavern floor on which

Borexino sits at a more or less constant temperature of 8 degrees. This, it turns out, is cooler than the average temperature 17 metres higher up at the top of the tank. Since hot things go up and cool things go down, the consequence is that the liquid inside Borexino remains almost completely still. Crucially this keeps any radioactive contamination that seeps out of the spherical nylon container inside the tank in place, stopping it from mixing with the liquid used to detect the neutrinos. The job of the new insulation and blue water pipes is to maintain as steady a temperature as possible, which in turn will keep the liquid inside from flowing about, which maybe, just maybe, will give Borexino a fighting chance of spotting the last missing reaction that the Sun uses to make helium.

We climbed down a staircase to stand at the foot of the towering detector. It is by far the strangest observatory I have ever seen, a silent giant, deep under a mountain, that waits patiently for whispers from ghostly emissaries from the centre of our Sun. As we left Hall C to return to the car, I asked Aldo what he thought the chances of seeing the CNO cycle were before Borexino's life came to an end. He looked at me sideways. 'We'll see . . . I would say, by the end of the year.'

A fortnight before my trip to Italy I'd had a Skype call with Gianpaolo Bellini, known affectionately as the 'father of Borexino', from his holiday villa in rural Italy. Although now in his eighties and retired (at least officially) he still radiates enthusiasm and delight in the success of the experiment that he first imagined in the early 1990s. Finally snaring the CNO cycle would be both a sweet end to a long career and a fitting reward for the hundred-strong team who have worked tirelessly to perfect their observatory. Once Borexino goes offline in two years' time, there are no new detectors in the works with similar capabilities. If Borexino doesn't see the CNO cycle, perhaps no one will.

That evening at my hotel, I thought about taking a drive up into the mountains to catch the sunset, but in the end I decided, sod it, I was tired and it was time for a beer and a pizza. You've seen one sunset you've seen them all, and in any case, unlike sunshine, neutrinos don't get blocked when the Sun goes down; they stream right through the Earth just as strong as ever. So, as I sat on the terrace of my hotel in the dwindling light sipping my beer I tried to imagine an invisible flood of neutrinos, trillions upon trillions strong, streaming through my body, a short moment on their long journey from the centre of the Sun out into the depths of space.

CHAPTER 6

Starstuff

It's several months now since I first caught the train down to my parents' house on a mission to break a Mr Kipling Bramley apple pie down into its chemical elements. I keep the products in sealed test tubes on my desk at home as a reminder that no matter how deep into the strange and abstract world of particle physics we get, we are still ultimately chasing the origins of ordinary stuff. Also, they're fun to look at.

The lump of carbon, the charred remnants of the apple pie, is my favourite: hard, jagged, and jet black, with small reflective faces that catch the light. Of all the elements, carbon has got to be the most charismatic. Its many and varied personas, from charcoal to diamond, make it the David Bowie of the periodic table, but its real mystique comes from its role as the key building block of life. All living things, from apple trees to Mr Kipling himself,* are built up from molecules with backbones of carbon.

The atoms that make up that little lump of carbon are truly ancient. They were forged long, long ago, before the first living thing, before the Earth formed, even before the Sun first flickered

---

* Or so I thought. Turns out that Mr Kipling never actually existed. He's a fake, a fraud, like the Wizard of Oz or Ronald McDonald, invented by brand consultants in the 1960s to flog baked goods. That said, branding consultants are still carbon-based life forms (my brother is one) so the point stands.

into light, somewhere, out there, in a distant part of the cosmos. The question is, where?

We have taken the first step towards a recipe for the elements in an apple pie. Thanks to more than a century of work by astronomers and physicists, we know that stars like our Sun are giant thermonuclear ovens that cook helium from hydrogen over billions of years. If the stars can make helium from hydrogen, then perhaps they can also make the heavier elements. Since an ordinary carbon nucleus is made of six protons and six neutrons, making one should be a simple matter of melding three helium nuclei together.

In fact, Hans Bethe made exactly that suggestion when he published his famous paper on why the stars shine way back in 1939. However, he ran into the same problem that Arthur Eddington had encountered when he first put forward the idea that the Sun might be powered by fusing hydrogen into helium – the stars didn't seem to be hot enough. As we saw in the last chapter, getting two protons to fuse meant finding a way to overcome the gigantic electrical repulsion between the two positively charged particles. The solution in that case was provided by Gamow, Houtermans, and Atkinson, when they showed that quantum mechanics allowed a proton to quantum tunnel through the walls surrounding the nuclear fortress, allowing fusion to take place at temperatures found in the cores of the Sun and stars.

Unfortunately, forcing three helium nuclei together to make carbon is far more challenging. A helium nucleus has an electric charge of +2, which means that the repulsive force between three helium nuclei is far stronger than for hydrogen. Hans Bethe realized that to get helium to fuse would require temperatures of hundreds of millions, perhaps billions, of degrees, far hotter than anyone imagined could possibly exist inside a star.

But if stellar ovens weren't hot enough to cook the heavy elements, then where do they come from? One daring solution was proposed by George Gamow and his PhD student Ralph Alpher in

1948. If the elements beyond helium weren't made in the stars, there was only one other furnace that could conceivably be hot enough – the primordial fireball at the dawn of time.

The idea that the universe had begun with a big bang had been gaining support since the 1920s, when astronomers had discovered that the universe seemed to be expanding. Winding the clock back, this implied that in the past the universe must have been smaller, and if you went back far enough you found that everything would once have been squashed into a single point.

According to Gamow and Alpher's theory, billions of years ago the entire universe was concentrated into a tiny, unimaginably hot embryonic blob, filled with a superheated gas of neutrons. For reasons unknown, that blob started to expand rapidly, and as it grew and cooled, the neutrons successively bumped into one another, building up the elements one by one in a frenzy of nuclear reactions, starting with hydrogen and running all the way through the periodic table.

Unfortunately, it soon became clear that the theory had a fatal flaw. Nature had rather unobligingly failed to include a chemical element with mass 5. That meant once you made helium (mass 4) the path was blocked. Adding another neutron to helium results in a nucleus so fantastically unstable that it falls apart in around a billionth of a trillionth of a second, far, far too short a time for another neutron to have a chance of smacking into it and getting you to mass 6. One way over the chasm at mass 5 might be to bang two helium nuclei together to make a nucleus of mass 8, but again the resulting nucleus's life is so fleeting – one ten-thousandth of a trillionth of a second – that there just isn't enough time to let you leapfrog to the next stable element at mass 9.

Gamow and Alpher's big bang had crashed and burned. And yet, we live in a universe with carbon, oxygen, iron, and uranium. They must have come from somewhere. But where?

Fortunately, at the very time that Gamow and Alpher were working on their theory in the United States, a young theoretical physicist from England named Fred Hoyle was puzzling over the same problem.

THE RECIPE FOR CARBON

Fred Hoyle was one of the most influential and controversial astronomers of the twentieth century. Born in Yorkshire in the north of England to a poor family in the wool trade, he had often played truant from school but caught the science bug when he borrowed a book from his local library by Arthur Eddington titled *Stars and Atoms*, two subjects that would come to dominate his life. Thanks in large part to the dogged efforts of a devoted teacher, Fred secured a scholarship to study at Cambridge, and by cock-up more than conspiracy, wound up as the PhD student of Paul Dirac, the greatest quantum physicist in the known universe. In the mid-1930s, Dirac, who was an otherwise rather unwilling supervisor, gave Hoyle one life-changing piece of advice: the glory days of physics were over, the quantum revolution was done, and the time wasn't yet ripe for new breakthroughs. If the ambitious young man wanted to make his mark on science, he should look elsewhere. So Fred Hoyle turned his attention to the stars.

During his long and eclectic career, Hoyle became famous for his contrarian, sometimes wacky scientific views, bitter disputes with fellow academics, and as a talented writer of science fiction, including the smash-hit BBC television series *A for Andromeda*. Today, he's probably best known as a die-hard opponent of the big bang theory, which he dismissed as pseudoscience thanks to its inability to explain what actually caused it to happen in the first place.*

---

* Hoyle is credited with coming up with the phrase 'big bang' in a 1949 BBC radio interview. Some claimed the term was meant as an insult, although Hoyle insisted that he was just trying to conjure a striking image.

Despite all the sound and fury that Hoyle generated, he was unquestionably a brilliant scientist. In fact, a key ingredient of his success was the same willingness to go against orthodox thinking that often landed him in hot water. As far as Hoyle was concerned, 'It's better to be interesting and wrong than boring and right.' One subject on which he turned out to be both interesting *and* right was the origins of the chemical elements.

In late 1944, Hoyle got a chance to leave the gloom of wartime Britain to visit the United States for a meeting on radar technology. While stateside, he took the opportunity to drop in on the Mount Wilson Observatory in California, where he cadged a lift back to Pasadena from the greatest observational astronomer of his age, Walter Baade. Baade and Hoyle soon got chatting about the most energetic outbursts in the known universe: supernovae. These stupendously violent stellar explosions can briefly blast out more power than all of the hundreds of billions of stars in a galaxy combined. At the time, no one could explain the source of a supernova's awesome power, but Hoyle's mind was set whirring when he bumped into a former colleague as he waited for his flight home in Montreal.

The colleague was Maurice Pryce, a British physicist who had mysteriously disappeared from the radar establishment in Portsmouth where Hoyle was based earlier that year. Although what exactly Pryce and his colleagues were doing in Montreal was a closely guarded secret, the presence of key nuclear physicists at the nearby Chalk River site left Hoyle in little doubt: they were working on an atomic bomb.

During his conversations Hoyle picked up a hint of the problem they were struggling with. The objective was to design a bomb that used the radioactive isotope plutonium-239 as the nuclear explosive. To trigger a nuclear chain reaction the scientists and engineers at Chalk River were trying to find a way to crush a sphere of plutonium in on itself, in other words to create an *implosion*. If the plutonium could be imploded fast enough, then a runaway nuclear reaction would be set up, releasing a vast amount of destructive power in a far larger *explosion.**

Back in Britain, Hoyle began to wonder whether a similar process might be responsible for supernovae. As a star ages, it gradually burns through its hydrogen fuel supplies until at last its entire core has been transformed into helium. Hoyle realized that robbed of the source of heat that keeps the star inflated, the core would start to collapse under the crushing force of its own gravity. As it implodes, this vast well of gravitational energy would be converted into heat, sending the temperature at the centre of the star skyrocketing and ultimately resulting in an unimaginably violent explosion – a supernova. A year or so later, Hoyle calculated that a collapsing star could in principle generate temperatures of more than 4 billion degrees Celsius, hundreds of times higher than anyone had previously imagined possible. Maybe, just maybe, stars were the cosmic cookers that made the chemical elements after all.

When he was released from his wartime work in 1945, Hoyle returned to Cambridge and to the world of astronomy. Although he had already put forward a theory of how the elements might be made in a collapsing star, the details were far from worked out, and without a specific recipe to get from helium to the heavier elements, it was doomed to fail just like Gamow and Alpher's.

Going back to Bethe's seminal paper on nuclear fusion from 1939, Hoyle hit upon the idea of trying to revive what became known as the 'triple-alpha process', the reaction where three helium nuclei combined to form a nucleus of carbon-12. Bethe had originally dismissed the reaction for two reasons: first, it required stupendously high temperatures beyond what was possible in any star,

---

* A plutonium implosion bomb was successfully tested on 16 July 1945 in the New Mexico desert. Another plutonium device was detonated over the Japanese city of Nagasaki a few weeks later on 9 August, killing between 39,000 and 80,000 people.

and second, the chances of three helium nuclei smacking into one another at the same instant were vanishingly small.

Hoyle's work on collapsing stars had already removed the first objection, and he thought he could spy a way of overcoming the second as well. Instead of three helium nuclei running into one another at the same time, what if two collided first to form a nucleus of beryllium-8 (four protons, four neutrons), before being hit by a third helium to make carbon-12? But hang on, didn't we already say that there's no stable element of mass 8? That's true; beryllium-8 only lives for one ten-thousandth of a trillionth of a second before splitting apart into two helium nuclei. But Hoyle realized this might not be an insurmountable problem. In a star with the right temperatures and densities, so many helium nuclei would be crashing into one another that enough beryllium-8 would get made to make up for the fact that it was constantly falling apart. This balance of creation and destruction would mean there was always a stable concentration of beryllium-8 in the star's interior, even though an individual nucleus lived such a fleeting existence.

The question was, would there be enough beryllium-8 around at any one time so that a decent number would get hit by another helium to make carbon-12? In 1949, Hoyle gave the problem to one of his PhD students. If the answer came out the way he hoped, the door would be open to fuse all of the elements in the periodic table.

Giving that particular problem to that particular student turned out to be a big mistake. About two-thirds of the way through, the student got fed up and threw in the towel. Having myself ridden the three-year rollercoaster of loneliness, confusion, and frustration that is the modern PhD, during which a friend and I frequently fantasized about running off to start a bakery, I can totally sympathize. The problem was that, according to Cambridge's rather archaic rules, Hoyle had gifted the problem to the student irrevocably and couldn't work on it himself unless the student cancelled his PhD registration.

Unfortunately for Hoyle, 8,600 kilometres away in sunny

California, Hans Bethe's young postdoc Ed Salpeter was thinking about the very same problem. Salpeter had taken a sabbatical from Cornell in upstate New York to spend the summer of 1951 working at the California Institute of Technology's Kellogg Radiation Lab with Willy Fowler, a burly extrovert from Ohio. Fowler had already made his name as the father of experimental astrophysics by using particle accelerators to recreate the reactions that power the Sun and stars in the lab. Salpeter, on the other hand, was a theorist in desperate need of data. In particular, he needed to know the precise energy of the beryllium-8 nucleus. He couldn't have come to a better place.

As luck would have it, Fowler's team had already made just the measurement Salpeter needed. Shortly after the war they had used their proton accelerator to knock chunks out of beryllium-9 nuclei, briefly creating beryllium-8, which immediately fell apart into two helium nuclei. Adding up the energies of the two helium nuclei that came whizzing out of the collision, Fowler was able to make an accurate measurement of beryllium-8's energy. To Salpeter's delight, it had almost precisely the right value to boost massively the rate of the triple-alpha process. What's more, it became clear to Salpeter that the fusion of helium into carbon-12 could take place at much lower temperatures than previously thought, a couple of hundred million degrees instead of billions of degrees.

Back in Cambridge, Hoyle read Salpeter's paper on helium fusion with growing frustration. You can almost picture him banging his fist on his desk and cursing Cambridge's out-of-date regulations. He had been well and truly gazumped by the dynamic duo of Salpeter and Fowler. However, rather than throw his hands up in despair, Hoyle channelled his anger into a renewed determination that would soon bear fruit.

In late 1952, he got an invitation to spend the following spring lecturing at the California Institute of Technology (Caltech). Swapping the gloom and rationing of post-war Britain for the sun-drenched orange groves of Southern California was an enticing prospect, and what's more Hoyle had already got a taste for the good life during his trip to the United States in 1944. While preparing for his lectures Hoyle went back over Salpeter's reaction and began to realize that there was something seriously wrong.

Once carbon-12 had been made inside a star, it would almost immediately smack into another helium nucleus to create oxygen-16. That's not a problem in itself – after all, oxygen is an important ingredient of the universe – but the trouble was that the reaction would go so fast that almost no carbon would be left over to make living things (or indeed apple pies). The fact that Hoyle, a carbon-based life form, was able to worry about such matters suggested that there must be some other process at work that stopped all the carbon from being burned up.

The solution that Hoyle hit upon was both brilliant and fantastically audacious. He realized that carbon-12 could only have formed inside stars if its nucleus had a very specific property.

Now, just like electrons in atoms, protons and neutrons inside atomic nuclei can exist in a wide variety of different states, known as energy levels. You can think of these energy levels as like the rooms in a large multistorey hotel. When the nucleus is in its lowest energy state, the protons and neutrons fill up the rooms nearest the ground floor, only taking up residence in the higher floors if there are no spaces left downstairs. However, if you smack a nucleus hard, maybe by firing a gamma ray at it, the protons and neutrons get thrown into an excited state – perhaps in the analogy there's a fire in reception and all the residents run upstairs in a mad panic and end up in rooms on higher floors. Anyway, in a nucleus, there is a well-defined set of these excited states, which are determined by the forces between the protons and neutrons and the laws of quantum mechanics.

Hoyle realized that if there was an excited state in carbon-12 with the same energy as a typical collision between a beryllium-8 and a helium nucleus inside a star, then the rate of carbon-12 production would get a big boost, more than compensating for the later reaction to make oxygen-16. He was even able to calculate the energy of this special state – it needed to be very close to 7.65 MeV.*

By the time Hoyle arrived at Caltech, he was itching to talk to Willy Fowler about his special carbon state. But when he tried to buttonhole him at a cocktail party arranged in Hoyle's honour, Fowler refused to talk shop, leaving Hoyle to make small talk with the rest of the Caltech faculty. However, Hoyle was a man on a mission, and the next day he burst into Fowler's office without so much as a by-your-leave, demanding that they drop what they were doing and use their particle accelerator to look for his predicted excited state.

Fowler was sceptical to say the least. Here was a funny little man with a strange accent making wild claims that he could predict energy levels in atomic nuclei, a feat that even the best nuclear theorists of the time couldn't pull off. Hoyle's claim was clearly ludicrous, he obviously knew nothing about nuclear physics, and, besides, they had already measured the energy levels of carbon-12 and found no sign of the state that Hoyle appeared to be obsessed with. Fowler gave him the brush-off, but Hoyle just would not let it rest, eventually managing to peel off one of the junior postdocs, Ward Whaling, and persuade him it was worth taking a second look.

Doing the experiment was a serious undertaking. Aside from the usual technical challenges that go along with creating nuclear reactions in a lab, just to get the experiment set up, Whaling and his colleagues had to manoeuvre a spectrometer weighing several tonnes down a narrow corridor, rolling it along on a bed made up of hundreds of tennis balls, while a group of undergraduates frantically ferried the balls from the back to the front. The experiment itself took place in a dark half-basement of the Kellogg Lab, with Hoyle watching on anxiously amid electrical cables and whirring

---

* An MeV is a megaelectron volt, the energy that an electron gets when you zap it with a million volts.

machinery. He later wrote that he had felt like an accused criminal on trial, except unlike a criminal he didn't know whether he was innocent or guilty.

Days of nervous waiting passed without result, as Hoyle repeatedly descended into the hot, cramped basement, emerging with relief into the Californian air at the end of each day. He was acutely aware of how silly he would end up looking if he had set Whaling and his team on a wild goose chase. However, after about two weeks of painstaking work, Whaling gave him the extraordinary verdict: they had found Hoyle's excited state of carbon-12 exactly where he said it must be. Everyone, including Hoyle, was stunned. Fowler in particular, who had been extremely dubious about the pushy little man from England, was so blown away by Hoyle's achievement that he arranged to spend the following year across the pond working with him in Cambridge.

Hoyle returned home on a wave of euphoria. When Whaling published the results a few months later, he put Hoyle's name first on the paper, a remarkable tribute considering he hadn't actually got his hands dirty doing the experiment. Once he had come back down to earth, Hoyle was left in awe at the precarious state of affairs that made the existence of life in the universe possible. Apart from the special life-giving state of carbon-12, he realized that if oxygen-16 had had a similar state, with an energy of 7.19 MeV, then all the carbon produced inside a star would immediately get converted into oxygen. When he consulted oxygen-16's nuclear properties he found a state perilously close to the danger zone, at 7.12 MeV. Likewise, if beryllium-8 had been stable instead of immediately falling apart into two helium nuclei, then helium burning would be so violent that stars would blow themselves to smithereens long before they could fuse a decent quantity of carbon, or indeed any of the other heavy elements.

Life in the universe seems to be balanced on a knife edge. Shift any of the states in beryllium, carbon, or oxygen just a whisker in the wrong direction and you end up with a carbon-free universe, one with no life, or at least not life as we know it. It's as if some great cosmic tinkerer has carefully arranged their subtle nuclear properties so that enough of these atoms could get forged inside stars, sprayed out across the cosmos, and then, by a series of random accidents over billions of years, come together to form walking, talking collections of atoms that spend at least some of their time wondering about how they got there. Nuclear physics, in other words, seems to be fine-tuned for life.

If you find all this a bit unsettling, you're in good company. Fine-tuning is one of the most controversial topics in modern physics, and it's not difficult to see why. Once you accept the premise you're led almost inevitably to some pretty non-scientific ideas: gods, multiverses, giant cosmic simulations, and more besides. (This whole issue will come back with a vengeance later on.)

Putting any existential angst aside for now, we've reached a big moment on our quest to make an apple pie from scratch. At long last we've found the recipes for two of the main products of my garage experimentation. First of all:

THE RECIPE FOR CARBON –
THE TRIPLE-ALPHA PROCESS

Step 1: Deep inside a star, smack two helium nuclei together to form a highly unstable beryllium-8 nucleus.

Step 2: Quickly now, and by quickly I mean in around one ten-thousandth of a trillionth of a second, fire in another helium nucleus and cross your fingers.

Step 3: If you're very lucky, that helium nucleus will fuse with the beryllium-8 before it can spontaneously disintegrate, producing a nucleus of carbon-12 in Fred Hoyle's special excited state.

Step 4: Time to cross your fingers again. Some of the
time, that excited carbon-12 nucleus will just
fall apart again, leaving the three helium nuclei
you started with. But with a bit more luck, the
excited state will instead de-excite by firing out
two gamma rays, leaving us with a newly
minted nucleus of good old carbon-12.

Using this recipe, we can leap across the yawning gaps in the periodic table at masses 5 and 8, taking us all the way from helium at mass 4 to carbon at mass 12. With that previously impassable chasm behind us, the way is open to fuse all the chemical elements from carbon to uranium. Just ahead of us is the next stop, oxygen-16, and the way to get there is remarkably straightforward:

THE RECIPE FOR OXYGEN –
THE ALPHA PROCESS

Step 1: Take a freshly baked carbon-12 nucleus and
smack it with a helium-4 nucleus.
Step 2: Voilà! Oxygen-16 (plus a bit of leftover nuclear
energy in the form of a gamma ray).

With these two recipes in hand, we can at last make two of the main ingredients of our apple pie. Of course we still haven't figured out the precise details of how and where these reactions actually happen. While Hoyle had good reason to suspect that carbon and oxygen were made inside stars that had exhausted their supply of hydrogen, the story of how and why this happens is complex, dramatic, and unerringly beautiful. And what's more, the stellar origins of the chemical elements are far from settled. Across the world, astronomers still continue to ponder and probe the deepest reaches of the cosmos in search of the stellar ovens in which the ingredients of our world were made.

THE LIVES OF THE STARS

Perched on a high outcrop of the Sacramento Mountains, amid fragrant pines and firs, are the white domes of the Apache Point Observatory. To the west, the ridge plunges through thick forest to the Tularosa Basin over a kilometre below and the dazzling gypsum dunes of the White Sands National Park. In the mid-nineteenth century this was the Old West of legend, where Apache tribes ruled over a wide fertile valley, until they were displaced by American cattle ranchers who overgrazed the land and turned it into an arid desert. Today, large tracts of this corner of New Mexico lie within a US military firing range, and across the mountains to the north-west is where the world's first atomic bomb was detonated in July 1945.

The telescopes of Apache Point are trained on far more distant and far more potent nuclear fires. From this commanding vantage point high in the mountains, astronomers survey light from hundreds of thousands of stars spread across the Milky Way in an attempt to unravel the evolutionary history of our galaxy and the origins of the chemical elements.

From my motel in Alamogordo, I had taken the road east, climbing from the desert floor into the mountains, past the cutesy village of Cloudcroft and onwards and upwards through woods of tall conifers. As I approached the observatory, a sign warned me that cars shouldn't drive up at night – the glare of headlights is the last thing astronomers need to contend with when stargazing.

It was mid-afternoon when I pulled up in front of the single-storey operations building. I was there to meet Karen Kinemuchi, one of Apache Point's professional observers who had generously agreed to let me accompany her during the coming night shift. I found her on the platform of the huge 2.5-metre Sloan Telescope, which hangs perilously over a sheer drop to the mountainside below, where she was debugging an electrical glitch with a colleague.

She greeted me with a smile and a handshake and gestured with undisguised pride towards the spectacular view over the basin to the San Andres Mountains beyond. It certainly was an incredible spot to do your day job. After a few moments soaking in the view, I made my opening conversational gambit, which, being British, was naturally about the weather. The Sun was shining, but there was a layer of hazy cloud to the south-west, which had been building through the afternoon despite forecasts for a sunny day and a clear night. Karen didn't seem too concerned though; the telescope scans the sky in infrared light and can see through this kind of cloud cover with ease as long as it doesn't get too thick. In any case, we could check the radar map when we got to the control room.

My visit to Apache Point had been inspired by a Skype call I'd had with another astronomer a few weeks earlier. Jennifer Johnson, professor of astronomy at Ohio State University, uses data from the Sloan Telescope to try to understand how different stellar processes forge the ninety or so naturally occurring elements in the periodic table. It's a story that's captivating and complex in equal parts, and despite all the progress since Ed Salpeter, Fred Hoyle, and Ward Whaling unlocked the recipe for carbon in the early 1950s, it's a story that's still being written.

Sitting in her office in Columbus, Ohio, surrounded by books and astronomical ephemera, Jennifer had cheerily talked me through our state-of-the-art understanding of where the chemical elements come from, often breaking into a smile or a laugh when she arrived at some particularly thorny challenge that was exercising her and her colleagues. The foundations of her subject, which is known as 'stellar nucleosynthesis' – the cooking of atomic nuclei in stars – can be traced all the way back to Hoyle's visit to Caltech in 1953. Bowled over by Hoyle's magician-like prediction about the origins of carbon, the nuclear physicist and head of the Kellogg Radiation Lab, Willy Fowler, had spent the following year in Cambridge. There he met the astronomical power couple Margaret and Geoffrey Burbidge, and with Hoyle they made up a formidable four-person team.

In 1957, their partnership resulted in one of the most significant papers in the history of astrophysics. Known colloquially as B²FH after its four authors, the paper is a nuclear cookbook, laying out an intricate web of reactions that could create almost every element in nature in a variety of different stellar ovens. However, the key difference between the stars and ordinary kitchen ovens is that their power comes from the nuclear cooking process itself, and it's the changing chemical composition of a star's interior that ultimately shapes its evolution, from its birth as a cloud of collapsing dust and gas to its spectacular death throes.

According to B²FH there is no single place in the universe where all the chemical elements are made. Instead there is a range of different stellar furnaces, each of which enriches interstellar space with different chemical elements; small stars like our Sun that die by slowly shedding their outer layers, giant stars that blow themselves apart in spectacular supernova explosions, and white dwarfs, dead stellar husks that can detonate violently when they gobble up too much gas from a companion.

Jennifer's mission is to try to weave together all these different processes to create a complete picture of the origins of the elements. In the course of her research, she has produced a beautiful colour-coded version of the periodic table, where each chemical element is shaded depending on where we currently think it comes from. The array of different hues scattered across the table, with many elements shaded in more than one colour, gives an impression of the long, varied, and interconnected evolutionary histories of the stuff from which we are made.

But before we get into some hard-core stellar physics, let's take a step back and consider how we know anything about the stars at all. In 1835, the French philosopher Auguste Comte declared that we would never know what stars were made from. Now, saying that we'll never know something is really just asking for trouble

– you can only ever be proved wrong – but on the other hand it wasn't an unreasonable statement given how stupendously far away the stars are. You can't exactly pop over to one and take a sample. But in just a couple of decades poor old Comte got oeuf all over his face thanks to the unexpected arrival of a revolutionary new technique: spectroscopy.

Spectroscopy emerged from the crucial discovery that different chemical elements absorb and emit specific colours, or more technically frequencies, of light. If you studied chemistry at school you may have got to throw powdered metals into a Bunsen flame, creating a brief, vivid burst of colour. Strontium, for example, turns the flame crimson, while copper produces a lurid green. The colour of a firework comes from the same effect. The set of frequencies that a given element absorbs and emits are specific to that element, representing a unique fingerprint that can be used to detect its presence in a Bunsen flame, a firework, or indeed in the fiery atmosphere of a distant star.*

If you take sunlight and break it into a rainbow spectrum using a prism and look very, very closely – you'll need to use a microscope – you'll discover that the rainbow band is littered with dark lines, not unlike a barcode. These dark lines correspond directly to chemical elements in the upper atmosphere of the Sun absorbing the light from the shining surface below. The discovery of spectroscopy was a revelation, transforming our understanding of the

---

* The reason why different chemical elements absorb and emit characteristic frequencies of light is all down to the quantum structure of atoms. As discussed in Chapter 3, electrons orbit the atomic nucleus in discrete, quantized energy levels, which are unique to each chemical element. When an electron makes a quantum jump to a different energy level it must either absorb or emit a photon, whose energy must be equal to the difference in energy between the two levels. To jump to a higher energy level, the electron must absorb a photon, and when falling to a lower energy level it emits a photon. Now the energy of a photon depends directly on its frequency (higher frequency = more energy) and a given atom will therefore absorb and emit photons of the specific frequencies that match the arrangement of its unique tower of energy levels.

heavens in a way not seen since the invention of the telescope at the start of the seventeenth century and heralding the birth of astrophysics.

The Sloan Telescope at Apache Point uses just this technique to decipher the code hidden in starlight and reveal the make-up of the stars in our galaxy. Later that afternoon as the Sun began to sink toward the San Andres Mountains, Karen pointed me to a small room just below the telescope platform that houses an instrument known as the Apache Point Galaxy Evolution Experiment, or APOGEE to its pals. Linked directly to the telescope above by bundles of fibre-optic cables, APOGEE can analyse light from a thousand targets simultaneously, allowing scientists like Jennifer back in Columbus to figure out what they are made from.

In the control room, surrounded by a bank of screens displaying the local weather map, live feeds from the telescope, and various graphs monitoring its performance, Karen talked me through what to expect on the night shift. The Sloan is operated by two astronomers, known as the warm observer and, rather ominously, the cold observer. The warm observer's job is to point the telescope at a list of target stars and galaxies while making sure that it's performing as expected, all from the relative comfort of the heated control room. Meanwhile, the less fortunate cold observer has to make multiple dashes into the freezing darkness to change 150-kilogram cartridges that plug directly into the base of the telescope, each containing a metal disc that acts as a star map, with hundreds of holes drilled at the locations of target stars or galaxies, connected to fibre-optic cables that carry the starlight to the APOGEE instrument.

Tonight, Karen had the good fortune of being the warm observer, but working the night shift is a gruelling task even if you're wrapped up indoors. She had only made it to bed after the previous shift at seven a.m. and had been up again by one p.m. to start running checks for the coming night. She wouldn't get back to her cabin for some more shut-eye until after sunrise. It was now late

November, and as winter drew on, the shifts would only get longer, darker, and colder.

Back outside, as we walked down towards the telescope, I made the mistake of trying to ask a slightly long and convoluted question and soon found myself gasping for air. Karen chuckled wryly as I recovered my composure; Apache Point is almost 3,000 metres above sea level, where the air is 25 per cent thinner than I'm used to. Up here you either walk or talk, but not both.

The Sun was now low above the distant mountains and there was a decided chill in the air. The clouds that had caused some concern earlier in the afternoon were dissolving, with the few remaining wisps glowing orange pink against the pale blue evening sky. It was shaping up to be a beautifully clear night, perfect conditions for stargazing.

After some checks of the telescope itself, it was time to open it up. With the push of a button and a short blast of a siren, the large white building that shelters the telescope began to slide back, running along a set of rails until the telescope was left alone and proud on the edge of the wide platform, nothing between it, the landscape, and the sky above. Then, slowly and silently, its huge barrel began to rise heavenwards, and in a moment of unexpected drama, the protective cover burst open like the petals of a flower opening to greet the Sun.

I found myself almost overwhelmed by the scene. The vastness of the landscape and the vivid colours above, which faded from orange through pink to deep blue and inky black, the brilliant diamonds of Venus and Jupiter chasing the Sun on its path towards the horizon, and beneath it all the silent telescope, thrust out into the cold, thin air, gazing upwards at the darkening sky. This has got to be science at its most romantic. Even Karen, who has watched this scene unfold countless times, told me that the magic never quite wears off.

As she headed back to the control room to prepare for the first observations, I stayed outside a little longer to watch the sunset.

Gradually the stars started to appear, one by one, each with its own story, its own past, and its own future. By human standards, the lives of stars are unimaginably long, most changing imperceptibly as the years roll by. Fortunately, the Milky Way provides us with hundreds of millions of stars to study, allowing astronomers to discern how they live and die from a myriad of stars at different points in their evolution.

A star's life is governed by the nuclear process occurring deep in its core. Take, for example, our Sun. As we've discovered, the Sun is currently fusing hydrogen together to make helium, and it will carry on doing this for another 5 billion years. However, this process cannot continue indefinitely. Slowly but surely, the Sun will burn through its hydrogen supplies as the core becomes increasingly rich in helium, which builds up like ash in a fireplace. Eventually, the Sun's hydrogen supply will run out, and when this happens things start to get interesting.

With its source of internal heat removed, the core will start to collapse in on itself under the pressure of gravity, heating up as it does so until it becomes so hot that it causes the burning hydrogen to flare up in a thin spherical layer surrounding the helium-rich core. This releases a flood of light into the overbearing gassy bulk of the star, blowing it up to monstrous proportions and turning our Sun into a swollen red giant.

This will be bad news for the Earth, which will most likely be engulfed in the Sun's scorching atmosphere.* Meanwhile, the inert helium core will continue to shrink and heat up, until it reaches a temperature of 100 million degrees, so hot that Salpeter and Hoyle's triple-alpha process can kick in and start fusing helium into carbon. This will result in a violent helium flash, releasing as much energy

---

* That said, life on Earth will have become pretty uncomfortable long before this – as the Sun ages it gets smaller and hotter, and in a mere billion years from now the Earth will become so hot that the oceans will boil, assuming we don't manage to do it ourselves first.

as the Sun radiates in 200 million years in about the same time as it takes to boil an egg.

Now burning helium in the core and hydrogen in a layer above, the Sun will shrink again by a factor of fifty until it's only ten times its current size, while it slowly manufactures carbon, some of which is then converted into oxygen by capturing another helium nucleus, producing two of the key ingredients in our apple pie.

However, this phase doesn't last very long, at least compared to the life of the star. After another 100 million years the helium too will run out, causing the core to resume its inward collapse, while helium and hydrogen continue to burn in concentric layers above. Mixing between these layers will also allow some of the carbon to fuse with hydrogen to make nitrogen (another element we'll need).

Now in its death throes, the Sun will undergo a final series of convulsions, gradually blowing its outer layers off into space and enriching the galaxy with carbon and nitrogen. Eventually, the last of its atmosphere will be wafted away, exposing a hot, dense core made almost entirely of carbon and oxygen – a white dwarf.

This is the end of the Sun. With the last nuclear reactions exhausted, all that remains is a glowing ember about the size of the Earth, surrounded by an expanding luminous cloud, the remnants of the Sun's atmosphere. The white dwarf itself is fantastically dense – a sugar-cube-sized lump would weigh around a tonne – and the only things that stop it collapsing any further are the laws of quantum mechanics, which forbid all its atoms from being in the same place at the same time.

We know all this thanks in large part to spectroscopic studies carried out with instruments like APOGEE. The light from the surface of a Sun-like star can reveal information about the process going on deep within, particularly later in its life when some of the products of nuclear fusion are dredged up to the surface by swirling convection currents. However, astronomers have also learned a lot from good old-fashioned observations using ordinary visible light.

Later that night, with the otherworldly glow of the Milky Way dominating the dark, moonless sky, I got a rare and completely unexpected opportunity to look through the largest instrument at Apache Point. Housed in a towering observatory building a short walk from the Sloan Telescope is the huge 3.5-metre ARC telescope. Normally it's controlled remotely over the Internet, allowing observers to study the sky from wherever they are in the world. However, tonight it was being fitted with an eyepiece so that a visiting group of University of Virginia PhD students could get some first-hand experience of stargazing. All was quiet in the control room, and so Karen suggested I tag along and take a look myself.

It was dark and freezing in the observatory, which was lit only by the stars shining through the narrow opening in the front of the building. ARC telescope observer Candace Gray drove the huge instrument from a computer at the back of the room. As she selected the first target for inspection, I felt the entire building begin to rotate beneath me and saw the stars move across the opening ahead, while the telescope pivoted to target the precise coordinates given to it by the computer.

To keep the students in suspense Candace gave them a clue to the first object of the night. 'Eleventh doctor', she said teasingly, but was met with dumbfounded silence. Clearly these American twenty-somethings weren't *Doctor Who* fans. 'Matt Smith, of course,' I said, rather pleased with myself. 'Bow ties!'

When my turn came, it took me a little while to get my eye adjusted, but when I did, a faint, delicate object came into view. I was looking at the remains of a Sun-like star, what astronomers call, rather misleadingly, a planetary nebula.* At the centre was a shining white dwarf, surrounded by two bulging lobes of luminous gas. If you used your imagination it did look a little bit like a rather

---

* The term 'planetary nebula' was coined in the late eighteenth century when astronomers didn't have much of a clue what they were looking at, thinking that they resembled fading planets.

badly tied bow tie, hence its nickname, the Bow Tie Nebula. I looked, transfixed, for a few moments – I had never seen a dead star with my own eyes before, and what's more here was exactly the type of object that produced most of the carbon in the world around us.

But what about the oxygen? Sun-like stars do indeed make oxygen at the end of their lives, but spectroscopic studies of planetary nebulae show that almost all of it remains locked up in the dense white dwarf, never escaping into the wider universe. Consulting Jennifer Johnson's colour-coded periodic table reveals that we must look elsewhere for the oxygen in our apple pies.

Back outside in the cold night air the Moon had risen, its light almost dazzling compared to the previous darkness. The Milky Way had faded out of view, with only the most brilliant stars still visible. Rising in the eastern sky was Orion, instantly recognizable thanks to its distinctive belt of three bright stars. According to the ancient Greeks, Orion was a hunter who got up to various shenanigans, including walking on water, drunken assaults on princesses, and threatening to kill every animal on Earth (he sounds like a swell guy) before finally coming a cropper fighting an oversized scorpion and getting shoved up in the sky by Zeus. Anyway, the stars in Orion are supposed to look like the figure of said hunter, if you really use your imagination and are a bit sloppy about the details.

Tracing your way from Orion's belt up to his left shoulder you'll find a particularly bright star with a distinctly reddish glow. Its name is Betelgeuse and it's an absolute monster, technically what astronomers call a red supergiant. If you were to pop Betelgeuse at the centre of our solar system, its vast, gassy bulk would engulf all of the inner planets, including the Earth and Mars, and stretch all the way out to the orbit of Jupiter. Betelgeuse is nearing the end of its life, a supersized vision of our Sun 5 billion years from now. However, its eventual fate will be rather more spectacular.

All other things being equal, the life of a star is dictated by its mass. The larger a star's mass, the more its core is crushed by gravity, and the more the core is crushed, the hotter it gets. As we've seen, at higher temperatures atomic nuclei whizz about at faster speeds, which means they can more easily overcome their electrical repulsion and fuse together. All this means that a heavier star burns through its nuclear fuel faster than a lighter star, and as the saying goes, 'The star that shines twice as bright shines half as long.' The Sun is relatively small in stellar terms and so will take about 10 billion years in total to burn through its hydrogen supply. Betelgeuse, on the other hand, is between ten and twenty times more massive than the Sun, and despite only being around 8 million years old has already consumed its hydrogen, like some enormous, voracious toddler, and swollen up to become a red supergiant.

We can't know for certain, but at best astronomers give Betelgeuse just another million years before all its helium is used up too, leaving a carbon–oxygen core. However, while this would be the end of the road for the Sun, Betelgeuse's giant size allows something extraordinary to happen.

Once helium burning ceases, the carbon–oxygen core will start to collapse under its colossal weight, heating up to more than half a billion degrees. At these ferocious temperatures, carbon nuclei are moving about so fast that they can overcome their tremendous electrical repulsion and fuse together to make even heavier elements, including neon, magnesium, sodium, and oxygen.

This phase of carbon burning lasts a mere one thousand years, a blink of an eye in stellar terms. Once the carbon runs out, the core goes through a series of further collapses, heating up each time and igniting new nuclear fuels, first neon and then oxygen. During this short period the star resembles a thermonuclear onion, with concentric layers fusing increasingly heavy elements as you move towards the core. In the star's last gasp, the core heats to a blistering 3 billion degrees, igniting the final nuclear reaction: fusing silicon into iron and nickel. This lasts for just a day.

Once the star's core is converted into iron and nickel, it's game over. These are the most stable nuclei in the periodic table, which means that fusing nickel and iron to make heavier elements actually uses up energy. The star has well and truly run out of juice. With no source of heat to fight against gravity, the core begins a final, unavoidable, and catastrophic collapse.

The core implodes, getting denser and denser as it falls inexorably towards oblivion. Nuclei are forced together until the entire heart of the star reaches the same density as an atomic nucleus. Now protons and neutrons don't like to be closer together than in an atomic nucleus, and so when this happens, the strong nuclear force fights back and the infalling matter effectively bounces, sending a cataclysmic shockwave tearing back upwards through the star. At the same time, electrons and protons are forced together to make neutrons, releasing a colossal wave of neutrinos so intense that it actually blasts the infalling bulk of the star back outwards into space.

The consequence is one of the most powerful events in the universe, a supernova. As the star is torn apart, it briefly pumps more power into space than all the hundreds of billions of stars in a galaxy. When Betelgeuse goes supernova sometime in the next million years, it will outshine the full Moon and will be easily visible in daylight.* Fortunately, Betelgeuse is far enough away from the Earth not to pose any serious risk, but it certainly will be one hell of a show. Also, Orion will lose his left shoulder, though he probably deserves it.

Supernovae play a pivotal role in creating the elements crucial for the existence of life. The oxygen, sodium, magnesium, and iron in our apple pie were forged billions of years ago in the cataclysmic deaths of giant stars. Their violent ends enriched the universe in heavy elements, which mixed together with the remnants of smaller

---

* Shortly after my trip to Apache Point there was a frenzy of speculation that Betelgeuse might be about to blow after it dimmed unexpectedly over the winter of 2019–20, but that was eventually put down to dust blocking its light.

stars like our own Sun and eventually formed the planet on which we live. Carl Sagan, who had an almost unrivalled ability to convey science at its most lyrical, put it beautifully: 'The nitrogen in our DNA, the calcium in our teeth, the iron in our blood, the carbon in our apple pies were made in the interiors of collapsing stars. We are made of starstuff.'

While a lot of the story I've just shared was laid out in the B^{2}FH paper of 1957, there is still much we don't understand about the origins of the chemical elements. 'Sodium is a complete disaster,' Jennifer told me during our Skype call, 'and we don't know who to blame!' Theorists used to think it was all made in supernovae, but the trouble is that supernovae should make magnesium and sodium together, meaning the amounts you see across the galaxy should be closely related. Curiously, Jennifer and her colleagues don't see as tight a correlation between these two elements as you'd expect, which seems to imply that at least some sodium must be made somewhere else.

However, perhaps the biggest challenge to how we think about the origin of the elements came just a few years ago. On 17 August 2017, the LIGO collaboration* – a pair of observatories situated 3,000 kilometres apart, in Washington State and Louisiana – detected gravitational waves produced by an almighty smash-up between two ultradense objects known as neutron stars. Admittedly, there's rather a lot to unpack in that sentence. What's a gravitational wave? you might reasonably ask. More on that to come – gravitational waves are just too big and important for an aside – but in short, they are ripples in the fabric of space and time produced when extremely massive objects bang into each other.

A neutron star, on the other hand, is a possible end result of a supernova. If the dying star is heavy but not too heavy (between around eight and twenty-nine Suns) then as the core collapses electrons will get forced into atomic nuclei, converting all the

---

* LIGO stands for **L**aser **I**nterferometer **G**ravitational-**W**ave **O**bservatory.

protons into neutrons in the process and ultimately resulting in a single, enormous atomic nucleus made entirely of neutrons. When the supernova has ejected the rest of the star into space, what's left is a tiny, incredibly dense neutron star with a mass of between one and two Suns but only around 20 kilometres across. If you thought a white dwarf was dense, half a cup of neutron star matter would weigh as much as Mount Everest.

If two of these neutron stars are produced close enough together, they can end up colliding, sending violent ripples out through space and time. When LIGO first detected one of these signals, astronomical observatories around the world swung their telescopes to point at the patch of sky where the waves appeared to have come from. Extraordinarily, they saw light, and when they analysed that light spectroscopically, they found evidence of vast quantities of heavy metals being created, from gold through to uranium. In fact, by one estimate the collision produced enough gold to make thirty solid-gold planet Earths. But before you get on the phone to Elon Musk with a get-rich-quick scheme you should know that the collision happened in a galaxy around 130 million light-years away, a long ride in even the zippiest of rockets.

For decades it was thought that the heavy elements in our universe were made in supernova explosions, but Jennifer and her colleagues now suspect that a large fraction came from these cataclysmic neutron star collisions instead. It's extraordinary to think that most of the gold in an ordinary piece of jewellery is really a little bit of a neutron star. (Admittedly there isn't much gold in an apple pie, but perhaps we can make it a fancy one with a bit of edible gold leaf on top.)

At Apache Point, Karen laboured tirelessly through the night, working her way through target after target, all the while making sure the Sloan Telescope and APOGEE were functioning as expected. Each time they had collected enough light for a given patch of sky, Viktor, the cold observer, had to trudge out into the pitch black, armed only with a small flashlight, to unload the spent 150-kilo cartridge and install the next one. By one a.m. I was flagging and crept off to grab a few hours' sleep in the nearby dorm building before getting up again at five a.m. to see Karen close up for sunrise.

I found her, tea in hand, looking a little bleary-eyed but pleased with how the night had gone. Observing conditions had been more or less perfect and the Sloan had functioned flawlessly. Within just a few hours the night's observations would be available to the hundreds of scientists working on the Sloan Digital Sky Survey all around the world.

For Jennifer and her colleagues, the hunt is now on for the oldest stars in the universe. As the universe has evolved, stars have increasingly enriched interstellar space with what astronomers refer to as metals, by which they mean any element heavier than helium. As a result, younger stars tend to be rich in metals, while the oldest stars are metal poor. Using APOGEE to determine what elements are present in the atmosphere of a star therefore allows astronomers to infer their age. The dream would be to find a star that formed from the pristine, unpolluted gas that filled the universe before the first stars blinked into light.

Such a star, if one is ever found, would be an ancient relic from the birth of the universe, made of about 75 per cent hydrogen and 25 per cent helium. However, this in itself raises a question. Over the past 13 billion years or so, generations of stars have transformed just 2 per cent of the matter in the universe into heavier elements. If that's the case, and assuming that all matter began as the simplest element, hydrogen, then where did all the helium come from? It was a question that Fred Hoyle and his collaborators were unable to answer back in the 1950s.

Outside, in the cold, crisp air, a faint light was growing behind the wooded ridge to the east. Karen and I walked quietly down to the Sloan Telescope, which stood gazing at the final target of the night. As she closed up the telescope, I asked how she coped with so many sleepless nights, so far away from home. She turned and gestured at the view. Across the Tularosa Basin, the lights of Alamogordo twinkled softly in the morning air, and the tips of the San Andres Mountains were catching the first light of the rising Sun. 'This', she said, 'makes it worth it.'

CHAPTER 7

The Ultimate Cosmic Cooker

The atoms that make up our bodies were forged billions of years ago deep inside stars.

That has got to be the most poetic idea ever to come out of science. It connects our ordinary, humdrum lives to the cosmic. We, and everything we see around us, apple pies included, are part of the story of the lives and the deaths of stars. Unsurprisingly, the discovery of our celestial origins quickly caught the imaginations of artists, writers, and musicians. In 1969, Joni Mitchell wove the idea into her countercultural anthem 'Woodstock', which expresses a young generation's yearning to achieve a more perfect state of harmony with itself and with nature: 'We are stardust (billion year old carbon) / We are golden (caught in the devil's bargain) / And we've got to get ourselves / Back to the garden.' That and to get stoned to the eyeballs in a field. Of course, this does all raise another question, namely, where did the stuff that the stars were made of ultimately come from? To a certain extent, the answer is other stars, which died and blew their matter into space, which then got mixed up with other dust and gas to form yet more stars. But at a certain point this chain of logic must break down.

The fact that there is still a large amount of hydrogen in the universe implies one of two things. If the universe is infinitely old and stars are continuously turning hydrogen into heavy elements, then new hydrogen must somehow be being created to replenish what the stars use up. The alternative is that the universe is not infinitely old and star formation began at some point in the past, billions of years ago perhaps, but certainly not *infinity* ago.

So the question of where the matter that makes the stars comes from is thus inextricably mixed up with an even more profound question, arguably the most profound question scientists have ever asked: *Did* the universe begin?

While Joni Mitchell was singing about stardust, a long-fought, sometimes bitter argument over the origins (or not) of the universe was coming to an end. On one side were those who argued that the universe has always been here, and that despite all the dynamism that we see in the sky, at the largest scales the universe is ultimately unchanging and eternal, with neither beginning nor end. Chief among the proponents of this steady-state universe was habitual contrarian and architect of stellar nucleosynthesis Fred Hoyle.

Arrayed against Hoyle and his collaborators were the supporters of what Hoyle himself had labelled the 'big bang', who argued that the universe had been born billions of years ago, bursting into existence from a single point of unimaginable density and creating space, time, light, and matter in the process.

Hoyle *hated* the idea of the big bang. As far as he was concerned, it was unscientific, involving a moment of creation whose ultimate cause could never be probed scientifically. Worse still, as an avowed atheist, it had the unpleasant whiff of religion about it. Allow the universe to have a beginning and you open the door to all kinds of mystical nonsense about how that beginning came to pass.

However, as we'll soon see, both the big bang and the steady state involve moments of creation of one kind or another. The big bang gets creation done all in one go at the beginning of the universe. Meanwhile, the steady state requires an infinite number of microscopic moments of creation, with individual particles of matter continuously popping into existence throughout all of space and time.

I'm sure you know how this debate ends – after all, there's no sitcom called *The Steady State Theory* – but the discoveries that led to the big bang and the steady state, as well as the observations that finally saw the big bang win, are absolutely crucial to our quest for the origins of matter. From here on in, as we leave the chemical elements behind and venture deeper into the structure of matter, we will find that there is only one oven that matters, the one at the beginning of the universe.

THE UNIVERSE EXPLODES

A few years ago, I attended a particle physics conference just outside Melbourne, Australia. It was a bit of a sweet gig (as nobody says in academia): the venue was in the seaside resort of Torquay, a mecca for surfers and the gateway to the Great Ocean Road, an improbable 243-kilometre stretch of asphalt that snakes its way westwards past sheer limestone cliffs, long beaches of white sand, and lush rain forests. Before the week of intense PowerPoint presentations kicked off, I hired a car and spent a few days exploring the famous route, stopping off in pretty seaside towns along the way.

One night, I was driving back to my hostel after a slightly disappointing evening bobbing about on a lake on what had been advertised as a 'platypus watching tour', during which the platypuses had remained conspicuously absent. Emerging from the forest, the unlit road turned along the coast, back towards my hostel at Apollo Bay. It was a particularly clear night and as I was miles from the nearest town I decided to pull over and take a look at the night sky.

Switching off the headlights, I stepped out of the car and looked up. What I saw made my head swim. Above me, stretched out across the sky, was the Milky Way, thousands of stars shining more brilliantly than I had ever seen before. I was suddenly overcome by what felt like vertigo, and for a brief moment I lost my balance and reached out to steady myself against the roof of the car.

Having spent most of my life living in or close to big cities, I had only seen the faint trace of the Milky Way a handful of times, but here, on a moonless night, far from any sources of light pollution, it dominated the sky. Directly above me was the glowing bulge of the galactic core, wreathed in the immense shadow of the Great Rift, a colossal band of molecular dust that hangs like smoke against the galaxy behind. Just to one side were two luminous patches – the Large and Small Magellanic Clouds – dwarf galaxies in orbit around the much larger Milky Way. The night sky at my home in London is a two-dimensional thing, a dark sheet pierced by a few pricks of light, but this scene was so brilliant, so detailed, that for the very first time I felt as though I was looking at a huge three-dimensional object.

I think that moment was the first time that I felt awe in the true sense of the word: a mixture of wonder, delight, and fear. The way the galaxy loomed above left me feeling insignificant and yet exhilarated at the same time. The experience reminded me of the Total Perspective Vortex from Douglas Adams's *Hitchhiker's Guide to the Galaxy*, a torture device that drove its unfortunate victims mad by showing them the unfathomable vastness of the universe accompanied by a microscopic dot on a microscopic dot labelled 'You are here.'

Until the 1920s, most astronomers thought that the Milky Way was the entire universe: a gigantic island of stars, alone in the darkness. However, there was a debate, occasionally fierce, over whether spiral nebulae, faint whirlpool-like smudges scattered across the night sky, were clouds of dust and gas in the Milky Way, or perhaps their own island universes far beyond the borders of our galaxy. The problem was that there was no way to measure how far away they were, that is until a crucial breakthrough made by the pioneering American astronomer Henrietta Swan Leavitt.

In 1904 Leavitt discovered a number of faint stars in the Small

Magellanic Cloud whose brightness seemed to change over time. During the next few years she found hundreds more of these variable stars and by 1912 had noticed a clear relationship between their brightness and how fast they brightened and dimmed; the brighter the star was on average, the slower it pulsed.

Leavitt's law, as it became known, was the crucial clue that would allow astronomers to measure the distances to objects outside our local galactic neighbourhood for the first time. By taking the pulse of one of these variable stars, an astronomer can figure out how brightly it shines, and if you compare that to how bright it *appears* (more distant stars look dimmer than nearby ones), you can tell how far away it is.

Then in 1923, the American astronomer Edwin Hubble discovered a variable star in the largest spiral nebula in the night sky, Andromeda. Using Leavitt's law, he estimated that Andromeda was almost a million light-years from Earth,* a shockingly huge number considering that the size of the entire universe had recently been estimated at only a thousand light-years. At a stroke the cosmos had grown in size by a factor of a thousand.

In just a few short years, the way people imagined the universe was transformed. It became clear that spiral nebulae weren't clouds of dust and gas in the Milky Way but galaxies containing billions of stars that lay far beyond the edge of our own. The universe was suddenly a much larger place, but an even more significant discovery was yet to come.

A decade earlier, Vesto Slipher, who despite sounding like a character from *Star Wars* was actually a real-life astronomer at the Lowell Observatory in Arizona, had made the startling discovery that the Andromeda nebula appeared to be hurtling towards the Earth at a speed of around 300 kilometres per second. As he studied

---

* Today the figure is even larger, 2.5 million light-years. A light-year is the distance travelled by light in one year, roughly 9.5 trillion kilometres, which is more than sixty thousand times the distance from the Earth to the Sun.

other nebulae, he found that they all seemed to be moving, but most of them were actually moving away from the Earth, some at incredibly high speeds of more than 1,000 kilometres per second. At first Slipher tried to make sense of their motions by suggesting that the Milky Way itself might be drifting through space relative to the nebulae, but without knowing how far away they were it was impossible to draw firm conclusions.

Armed with Leavitt's law, Hubble was now ready to attack Slipher's puzzle. Working at the Mount Wilson Observatory in California, he made careful studies of variable stars in twenty-four galaxies outside the Milky Way and calculated their distances. Comparing his results with Slipher's estimate of their speeds he found an intriguing pattern. With the exception of the very closest galaxies to the Milky Way like Andromeda, every galaxy in the night sky appeared to be moving away from the Earth, and the farther away they were the faster they were retreating. It was as if the entire universe was expanding away from us, although when he published his results in 1929 Hubble himself was careful not to make such a bold claim.

At first some questioned whether Hubble's results were reliable, but by 1931 he had produced new measurements including galaxies more than 100 million light-years away. The new data left little room for doubt; the effect was real. What's more, there was an unmistakable linear relationship between the speed of a galaxy and its distance – in other words, a galaxy twice as far away from the Earth would be moving away twice as quickly. The controversial bit was how to interpret it. Many physicists, including Einstein, had been wedded to the idea of a static, unchanging, eternal universe. To admit that the universe might be expanding opened up the possibility that it might have had a beginning, and that idea made many physicists and astronomers feel unwell.

One man who felt no such queasiness was the Belgian physicist and Catholic priest Georges Lemaître. Not only did Lemaître argue that the universe was expanding, he took the argument to its logical extreme: if the universe was getting bigger, then in the past it must have been smaller, and if you keep winding the clock back then eventually you arrive at a point where everything in the universe was squashed into a single unimaginably dense object, what Lemaître called the 'primeval atom'.

Taking his inspiration from radioactivity, Lemaître imagined the primeval atom as an atomic nucleus, just a really, really heavy one weighing as much as the entire universe. According to Lemaître the universe began when his cosmic nucleus suddenly exploded like a firework, shattering into star-sized atoms, which carried on breaking apart into smaller and smaller pieces, eventually giving rise to everything we see around us.

But unlike a firework, Lemaître's primeval atom didn't explode into a space that already existed – the explosion *was* space. Before the firework, all of space was squashed down inside the primeval atom, and afterward, it was space itself that expanded. There was no centre to Lemaître's firework – the explosion happened everywhere in the universe at the same time. Everywhere was inside the primeval atom and the primeval atom was everywhere. It's true that almost every galaxy in the sky is rushing away from us, but at the same time they aren't actually moving through space. It's the space in between that's getting bigger, carrying every galaxy farther and farther apart as if they were sitting on the surface of some vast inflating balloon.

Although the firework universe made Lemaître something of a public celebrity, it's fair to say that his theory didn't meet with universal acclaim in the scientific community. Many were dismayed by the idea of a moment of creation, specifically the possible role it might leave for a creator. Arthur Eddington, who had a huge amount of respect for his talented former student, called the idea 'repugnant'. However, Einstein, who had been troubled for some time by his attempts to maintain a static universe, was won over, describing Lemaître's theory as 'the most beautiful and satisfactory explanation of creation to which I have ever listened'.

The trouble was that Lemaître's theory was just one of many possible ways of describing the evolution of the universe. By the 1930s, all such cosmological models were based on the powerful framework provided by Einstein's general theory of relativity. This jewel-like theory was Einstein's masterpiece, an elegant and highly mathematical reimagination of what we mean by space, time, and gravity. General relativity made it possible to write down a single equation describing the size, shape, and evolution of the entire universe, effectively creating the modern science of cosmology, the study of the universe as a whole. However, there wasn't one unique equation, there were many, each describing a different universe with a different history and a different future. Lemaître's firework universe was just one of these, and while it explained why almost every galaxy appears to be rushing away from us, there were other theories that could do the same thing without the need for a philosophically troubling moment of creation.

For the most part, such high-minded cosmological debates remained the preserve of devotees of general relativity, among them Einstein, Lemaître, and Eddington. However, in the late 1930s nuclear physicists started to take an interest in Lemaître's firework universe, not as a way of solving problems in astronomy per se, but as a way to unlock the origin of the chemical elements. At the time, the stars weren't thought to be hot enough to make the heavy elements, but there was one place that certainly would have been: the ultimate thermonuclear oven at the dawn of time.

FLASH-FRIED HELIUM

There isn't any one person to credit with the idea of the big bang. The theory came together slowly in fits and starts, the work of many hands. That said, one person who did a lot to get the whole thing going was George Gamow.

Gamow never planned to come up with a theory of how the

universe began. He was led to the big bang almost by accident during his quest to unlock the origins of the elements, a line of research that goes all the way back to his lively coffee-fuelled discussions with Fritz Houtermans in the summer of 1928. Gamow was well equipped to get to grips with the universe as a whole. As a young student in Petrograd he had learned general relativity from one of its all-time greats, the Russian physicist and mathematician Alexander Friedmann, who had been the first person to use Einstein's theory to write down an equation describing an expanding universe.

When Hans Bethe had argued in 1939 that the stars weren't hot enough to fuse elements beyond helium, Gamow began to wonder if the beginning of the expanding universe might have had the right conditions to do the job. Unlike many of his colleagues, Gamow found himself shut out from research on atomic bombs during the Second World War. Perhaps the US authorities were worried by his Russian roots, or maybe it was just his rambunctious personality and love of a good story told over a martini or three. In any case, he was left with plenty of time to develop his ideas, and by the end of the war he had the skeleton of a theory.

Gamow's big bang started with a universe that was incredibly small and dense, filled with a cold but fantastically thick soup of neutrons, packing a tonne into every cubic centimetre. The universe then expanded, following an equation discovered by Friedmann and Lemaître, increasing in size by a factor of ten in a little over a second. As space expanded, the neutrons fused together, or 'coagulated' as Gamow put it, to form larger nuclei made of neutrons. At the same time, some of the neutrons decayed into protons, converting these neutron-only nuclei into ones made of both protons and neutrons, and eventually all the familiar chemical elements.

Gamow's scheme was unbelievably ambitious: his goal was to make every single chemical element in one almighty explosion at the dawn of time. To have any chance of being right it would need to accurately reproduce the relative amounts of each element as we find them in the world around us. Fortunately, just before the war the Swiss-Norwegian geochemist Victor Goldschmidt had published the results of a wide-ranging survey of the abundances of the elements based on collated measurements of rocks, meteorites, and the spectra of starlight. Goldschmidt's data was an archaeological treasure trove capturing the entire cosmic history of the elements. If Gamow's big bang could replicate Goldschmidt's data, then he would be onto a winner.

Gamow may have had a fantastic imagination, but he didn't have much enthusiasm for the hard work of calculating the detailed consequences of his wild ideas. Instead, he gave the problem of predicting how much of each element should be made in his big bang to his PhD student, Ralph Alpher. Alpher was a bit of a novice when it came to nuclear physics but had been bounced into the area after he discovered to his dismay that the thesis topic he had spent a year working on had already been published by another physicist.* At the time, Gamow was out on a bit of a limb, and so at least, Alpher thought, he would be relatively free from competition.

During his PhD studies, Alpher was pulling a forty-hour week on military research at the Johns Hopkins Applied Physics Lab and moonlighting with Gamow on physics after hours at George Washington University. They would meet to discuss progress at Gamow's favourite Washington restaurant, Little Vienna, where Alpher would snatch a bite to eat between jobs while the hard-drinking Gamow knocked back martini after martini. Alpher was soon having more sober discussions with his neighbour at Johns Hopkins, Robert Herman, who was fascinated by Alpher and Gamow's theory and soon joined them as a collaborator.

The big breakthrough came when Alpher happened to hear a talk given by the physicist Donald Hughes about his experimental

---

* In fact, he was so pissed off that he literally tore up a year's worth of notes and flushed them down the toilet.

work firing neutrons at various elements. Hughes was interested in how different materials would fare in the harsh environment of a nuclear reactor and so had been bombarding as many elements with neutrons as he could get his hands on. Alpher immediately realized that Hughes's data was exactly what he needed.

Comparing Hughes's neutron data with Goldschmidt's element abundances, Alpher noticed a pattern. Elements that had a high tendency to gobble up neutrons tended to be rarer, and vice versa, which makes sense if you think about it a bit. An element that's good at swallowing neutrons would have been converted into heavier elements during Gamow's big bang, leaving relatively little of the original element. On the other hand, an element that's unlikely to absorb a neutron will tend to hang around once it's been made, leading to a higher abundance. It was only indicative, but it seemed to suggest that Gamow was onto something.

Alpher completed his PhD dissertation in the summer of 1948. Along the way, he and Herman had realized that Gamow's original assumption of a cold neutron soup was wrong. Instead, the early universe would have been dominated not by neutrons but by light and would have been so ferociously hot that any elements that formed in the first few minutes would be immediately smashed apart again by collisions with high-energy photons. In the refined model, the cosmic cooking process only kicked in when the universe had been expanding for around five minutes and cooled down to a billion degrees.

However, neutrons only live for an average of fifteen minutes before decaying into protons, electrons, and neutrinos, and so many of them would have died off during those first five minutes. As a result, the first nuclear reactions would have involved a proton and a neutron fusing together to make the heavy isotope of hydrogen known as deuterium. Once some deuterium had been cooked, it could then swallow another neutron or proton to make either tritium (a heavy isotope of hydrogen made of one proton and two neutrons) or helium-3 (two protons and one neutron). These could then fuse to make helium-4, which could then swallow more neutrons until eventually you'd made all the elements in the periodic table. To his delight, when Alpher calculated the abundances of the elements predicted by the big bang, his result matched Goldschmidt's data rather well.

When Alpher and Gamow published an outline of the theory in the spring of 1948 it caused a media sensation. An article written by a local journalist got syndicated all over the country, with the *Washington Post* reporting that 'The World Began in Five Minutes'. Gamow, in typical Gamow style, put it rather more colourfully: 'The elements were produced in less time than it takes to cook a dish of duck and roast potatoes.'

Several journalists seized on the symmetry between the creative thermonuclear explosion of the big bang and the destructive explosion of a nuclear weapon. Others strayed into religious territory, causing Alpher to receive several letters from concerned Christians offering prayers for his soul, even though he had been careful to avoid any mention of God. Such was the media frenzy that when Alpher made his oral defence of his PhD, around three hundred people crowded into the room at George Washington University to listen in on how the universe began.

Over the next few years, Alpher and Herman forged ahead, developing the big bang into a proper, quantitative scientific theory. However, the whole project soon hit the rocks. One rather embarrassing issue that had been bubbling away for almost two decades concerned the age of the universe. Cosmologists could use Hubble's measurement of how fast the universe was expanding to rewind the clock and figure out how far back in time the big bang was supposed to have happened. The answer came out at around 2 billion years, an inconveniently short figure given that radiometric dating suggested that the Earth was more than 4 billion years old. How could the universe be younger than the Earth?

Not a small problem, I'm sure you'll agree, but Gamow wasn't deterred. In 1949 he showed that a bit of judicious fiddling with the cosmological equations could lengthen the age of the universe as much as you liked. However, this involved some pretty shameless jiggery-pokery, which Einstein among others was seriously unhappy with.

An even more serious flaw in the Alpher–Gamow–Herman theory was the same issue that the stellar physicists had been struggling with: the fact that there were no stable nuclei with masses 5 or 8. Once you'd made helium-4 in the big bang you hit a dead end. Adding another neutron or fusing two helium nuclei together both got you nowhere. Alpher, Herman, and a number of other physicists including the great Italian Enrico Fermi tried in vain to find a route over the mass gap, but every time they thought they'd managed to erect a rickety bridge to the heavier elements, the whole thing came crashing down.

With the big bang theory seemingly on its last legs, one of its chief opponents, Fred Hoyle, was more than happy to put the boot in. Driven by a deep aversion to the idea of a moment of creation, he and his collaborators across the Atlantic in Cambridge, Hermann Bondi and Thomas Gold, had been developing a radical alternative history of the universe, the steady state, arguing that the universe has always been here, always will be, and despite the endless cycle of stellar birth and death, is unchanging.

The problem was, how can you have an expanding universe that always looks the same? The solution was hit upon by Gold – the spontaneous creation of matter. As the universe expands and the galaxies fly farther and farther apart, Gold suggested that atoms could be constantly popping into existence to fill in the gaps. This new matter would eventually clump together to form new stars and galaxies as the old ones age and fade away, keeping the universe looking the same indefinitely.

It was a pretty mad idea at first glance. For one thing, having matter appear out of nowhere violates the conservation of energy. But on the other hand, you only really needed a very small amount of matter creation to keep the universe steady, 'about one atom every century in a volume equal to the Empire State Building', as Hoyle vividly put it.

With Gamow and his colleagues struggling to 'make' the heavy elements in the big bang, the steady state won a major victory in 1957 when Hoyle, Fowler, and the Burbidges published their tour-de-force paper on how stars cook the chemical elements, the one familiarly known as B²FH. At a stroke, the big bang's original raison d'être had been shot to pieces. There was no need for a big bang to make the heavy elements; the stars could do it just fine on their own, thank you very much.

And yet, and yet. Even as the steady state seemed to be in the ascendence, portents of its downfall were beginning to appear in the heavens. Improved measurements of the expansion of space had been gradually lengthening the age of the universe, until by 1958 it had grown to as much as 13 billion years, far older than the oldest rocks found on Earth. Meanwhile new measurements of X-rays and radio waves from deep space challenged some of the basic tenets of the steady state, to the point that even some of its advocates began to abandon it.

Another problem that had been strangely ignored when B²FH came out was the thorny issue of helium. Helium is the second most abundant element in the universe, making up 25 per cent of the total mass of all atoms compared to 75 per cent for hydrogen and just a tiny sprinkling of heavier elements. Everything else – the carbon in our bones, the oxygen we breathe, the iron in our blood, the edible gold leaf on top of our apple pie – is just a light dusting of icing sugar on a vast hydrogen–helium cake. However, since stars make helium *and* all the other elements together, there is no way that they could have made such huge quantities of helium and yet so little of all the rest. Assuming that all matter started out as hydrogen, most of the helium in the universe must have come from somewhere else. But where?

Hoyle, whose faith in the steady state bordered on fanatical, tried to get around the problem by proposing the existence of gigantic 'black stars', enormous objects thousands or even millions of times heavier than the Sun, conveniently hidden from view inside giant clouds of gas. Thanks to their gigantic sizes, these megastars would go through a series of violent explosions and collapses, sort of like mini big bangs in their own right, generating temperatures of tens of billions of degrees in their cores and fusing vast amounts of hydrogen into helium. Unfortunately for Hoyle, there was zero evidence that these black stars really exist in the universe, and many of his colleagues saw it as a desperate attempt to salvage a dying theory.

On the other hand, it seemed that the big bang might fit the bill. Despite not being able to make the heavy elements, the theory had no trouble at all with making helium. The question was, how much helium would you expect to get fused in a hot big bang, and, crucially, does this match what we see in the universe around us?

To answer this question, we need to go right back to the first few minutes of the universe's history, to a time when all of space was filled with a blazing plasma of particles. Today, our universe is dominated by matter – gas, dust, stars, and dark matter* scattered across an otherwise empty void – but back then it was ruled by light. You might even say that the universe was made of light. The matter particles, the protons, neutrons, and electrons that would go on to make up everything we see around us, were little more than froth riding on a furious ocean of photons.

In the first few minutes, this primordial light was so fierce that a single photon carried enough energy to smash an atomic nucleus to pieces. As a result, almost no nuclei could form. If a proton and a neutron managed to fuse to make deuterium, then they were instantly blown apart again by a collision with a high-energy photon. However, in these first few minutes the universe was expanding extremely rapidly, and as it expanded, it cooled. After about three

---

* If you don't know what dark matter is, don't worry, neither do physicists. We'll come to that later . . .

minutes the cosmic oven had cooled to a few billion degrees and photons no longer had enough oomph to destroy a deuterium nucleus. Suddenly, the amount of deuterium in the universe sky-rocketed, and the cosmic cooking process roared into action.

In little more than a minute, a blizzard of nuclear reactions converted deuterium into tritium and helium-3, and then into helium-4. After about a hundred seconds, almost all the available neutrons had been consumed and it was all over. A few nuclear reactions carried on at a fairly desultory rate for the next little while, but just twenty minutes after the big bang the cosmic oven grew too cool, thermonuclear cooking came to an end, and the amount of helium in the universe was set.

But how much helium? Remarkably, the answer depends on a single ratio: the number of neutrons for every proton at the moment nuclear fusion gets going. Since almost all the neutrons end up getting converted into helium, and helium contains two neutrons and two protons, this simple ratio tells us precisely how much helium gets made. And the number of neutrons depends crucially on what happened in the very first second.

During the first second of cosmic time, the energy of particles in the primordial fireball is so high that neutrons and protons were continually being converted into each other by collisions with high-energy particles. At first, the reactions that convert protons to neutrons ran at the same rate as the opposite reactions converting neutrons back into protons. Equality reigned.

However, as the universe cooled, the fact that the neutron is slightly heavier than the proton began to upset the balance. The extra bit of energy needed to turn a proton into a neutron starts to make the reaction less likely than its reverse, and the number of neutrons compared to protons falls. When the first second is up, the temperature of the universe drops to the point where particles no longer have enough energy to turn a proton into a neutron and the number of neutrons freezes, with about one neutron for every six protons.

All we have to do now is wait a couple of minutes for the universe to cool down enough for nuclear fusion to get going. However, during that time, yet another factor comes into play – the fact that the neutron is unstable, only living for about fifteen minutes on average before it decays into a proton, an electron, and an antineutrino. As a result, during this two-minute wait, a decent fraction of the neutrons decay into protons, leaving just one neutron for every seven protons when fusion kicks off.

In the next minute or so almost all those neutrons get fused into helium-4, which, as we said, is made of two neutrons and two protons. So if we start out with fourteen protons for every two neutrons, then for every helium nucleus we expect twelve protons left over. Since helium weighs four times as much as a proton, that's a ratio of 4:12 helium to hydrogen. In other words, the big bang theory predicts that 25 per cent of the mass of atoms in the universe should end up as helium, with the remaining 75 per cent left over as unfused hydrogen. This is precisely what we see in the universe today!

Apologies for the mental arithmetic required there, but I hope the final message is clear enough: the big bang theory does a great job of predicting the amount of hydrogen and helium that astronomers see when they look outward into space. Hoyle himself came to the same conclusion in a paper he published with a younger colleague, Roger Tayler, in 1964. However, while Tayler took this as clear evidence for the big bang, the dogmatic Hoyle still wouldn't let go of the steady state, clinging stubbornly to his unseen dark stars.

By the mid-1960s the great war over the history of the universe was all but done.

The killer blow came in 1965 when two American radio astronomers, Arno Penzias and Robert Wilson, discovered a faint microwave glow emanating from the entire sky. The pair had been planning to use a large antenna at Bell Labs in New Jersey to study radio emissions from objects in the Milky Way. While they were calibrating their equipment, they were plagued by a low-level microwave noise that they just couldn't seem to get rid of. Realizing that the noise would make precise astronomical observations impossible, they spent the best part of a year trying to figure out where it was coming from.

They ruled out a catalogue of potential sources, both in space and on Earth, including stray radio broadcasts from New York City, just a few kilometres away across Lower Bay. In fact, no matter where they pointed the antenna, the noise remained stubbornly constant. It was only after much confusion and endless checks of the giant radio horn, which famously included evicting some roosting pigeons and cleaning up the 'white dielectric material' that they had left behind, that the significance of their discovery was realized. The faint microwave signal wasn't pigeon shit, it was the afterglow of creation.

Forgotten to almost everyone, back in 1948 Alpher and Herman had predicted that the fearsome light that had dominated the early fireball of the big bang should still be around today. Around 380,000 years after time zero the universe should have cooled down enough for negatively charged electrons to bind to positively charged nuclei and form the first neutral atoms. Before this pivotal moment in cosmic history, photons couldn't travel far through space without pinballing off charged particles in the primordial fireball. However, as the first neutral atoms formed, the universe changed from a fiery plasma to a transparent gas of hydrogen and helium. Suddenly, the photons were free to travel unimpeded through space.

That light has been travelling through the cosmos ever since. As it travelled, the expansion of space gradually stretched what started out as short wavelength visible light with a temperature of around 3,000 degrees to a weak microwave signal, just 2.7 degrees above absolute zero. It was this faint glow from the universe's fiery birth that Penzias and Wilson had stumbled upon. It appeared to come from the entire sky because the big bang happened everywhere, or to put it another way, everywhere was once inside that ancient fireball.

The discovery of what is now known as the 'cosmic microwave background' was the final piece of evidence that convinced cosmologists that our universe really did begin with a bang. I can't think of a scientific discovery more profound than that. In the space of just a few decades we'd gone from believing that our Milky Way galaxy was the entire universe to finding ourselves gazing out at an enormously vaster, ever-expanding cosmos whose origins can be traced back to a single unimaginably violent event that took place 13.8 billion years ago. Penzias and Wilson were rightly rewarded with a Nobel Prize for the painstaking work that had made the discovery possible, a great example of how scrupulous experimental care can lead to seriously big discoveries. As the science fiction writer Isaac Asimov once wrote, 'The most exciting phrase to hear in science, the one that heralds new discoveries, is not "Eureka!" (I found it!) but "That's funny . . ."'

Gamow, Alpher, and Herman, on the other hand, were left feeling more than a little bitter. Their original prediction of the cosmic microwave background had been all but ignored. It was the Princeton physicists Robert Dicke and Jim Peebles who'd realized the significance of Penzias and Wilson's microwave buzz, but when they published their paper, they were completely unaware that Gamow, Alpher, and Herman had made the same prediction almost two decades earlier. In fact, despite all their many contributions to our understanding of the origins of the elements, the stars, and the universe itself, neither George Gamow nor Fred Hoyle received a Nobel Prize. Perhaps Gamow's refusal to take anything seriously and his tendency for embarrassing drunken antics had played a part in him being passed over. In Hoyle's case, his downright rudeness and increasingly bonkers scientific views in later life had alienated many of his colleagues, views which included the idea that flu outbreaks were caused by microbes raining down from outer space and that the fossil of the birdlike dinosaur archaeopteryx was a fake.

Accolades or no, together Gamow and Hoyle laid the foundations of our understanding of where the chemical elements come from and, ironically, they were both right and wrong at the same time. The elements weren't all made in stars as Hoyle had earnestly hoped, nor were they all made in the fiery maelstrom of Gamow's big bang. They were made in both. The big bang gave birth to our universe, and in the process seeded space with the hydrogen and helium* that went on to form the first stars. These in their turn fused everything else, from the carbon in our apple pie to the uranium that warms our planet's core. We, and everything around us, are products of these awesome events. We are children of both the big bang and the stars.

We've reached a turning point in our cosmic cookery story. At last we know where the chemical elements that bubbled out of that first silly apple pie experiment came from. The carbon was made in stars like our Sun as they reached the end of their lives, the oxygen was blasted out across space by terrifying supernovae. The stars, in their turn, ultimately formed from the hydrogen and helium left over from the big bang. But there is one apple pie ingredient whose origin we still haven't explained, the simplest of all, the raw materials from which all the others got made: hydrogen.

In a sense, we already know where hydrogen comes from: the first hydrogen atoms formed 380,000 years after the big bang when protons and electrons got together for the first time. When I say we don't yet know where hydrogen came from, what I really mean is we don't know where *protons and electrons* came from. To answer that question, we must finally leave the chemical elements behind, delving into the wonderful world of particles, while drilling ever deeper into the very first second of cosmic history.

* And some tiny amounts of the third element, lithium.

CHAPTER 8

How to Cook a Proton

The first time I saw data from the Large Hadron Collider was a grey Friday morning in April 2010. I was at my desk in the bowels of the new Cavendish Laboratory, a heap of concrete drabness built in the 1970s after the famous lab had outgrown its creaking city-centre site and relocated to a windswept field on the edge of Cambridge.

I shared my windowless office with two other graduate students. One was a depressive Italian who spent most of his time bemoaning the backward state of British plumbing – 'Why don't you have mixed taps?' was a frequent refrain. 'When I wash my face I either freeze or I scald myself. It's not fit for human life . . .' – and a sardonic final-year student in the midst of writing up her dissertation, whose gallows humour made me and the Italian fear what ordeals still lay ahead.

I had just got back to Cambridge after spending the winter at CERN preparing for the first high-energy collisions at the LHC. Having lived the last few weeks in a state of mild terror at the prospect of being summoned to the control room to fix a problem that I had no idea how to solve, delays with the collider meant that I had left Geneva just as the first protons smashed into one another inside the LHCb detector. Almost as soon as the news arrived, so did an email from my supervisor asking whether I'd looked at the data yet.

The algorithm used to sift through the collision data in search of the specific particles we were interested in was already written and prepared. It was more or less a simple job of pressing go and waiting. The data had been steadily accumulating since the first collisions on 30 March, with each collision adding a little more to the small but rapidly growing store of new information about the subatomic world.

Today, running over the vast LHCb dataset takes weeks, but in those early days so few collisions had been recorded that I had the results a little more than an hour after setting the algorithm running. Opening up the data file I hurriedly navigated my way to examine the key graph that I knew would tell us whether or not we were in business.

A shaky double click on the mass spectrum, and there it was, a clear spike rising high above the low-level background noise, the unmistakable signature of the particle we were looking for. I remember feeling a rush of excitement. Up until that day I had only studied computer simulations, but here on my screen, clear as day, was proof that these particles actually did exist *in the real world*. Not only that, but what I was seeing was a product of the most ambitious scientific project ever attempted, a particle collider the size of a city that had taken decades to design and construct, an impossibly complex detector assembled by an international collaboration of more than seven hundred scientists, a worldwide grid of computer farms that stored, processed, and distributed the data across the globe, and, at the end of it all, the little algorithm that I had written. Somehow, miraculously, it had all worked.

I bashed out an excited email to my supervisor, Val Gibson, head of the Cambridge group, attaching a copy of the telltale graph. The spike was a clear sign that D mesons – exotic particles about twice as heavy as a proton – were being produced in the collisions in the heart of our detector. Seeing them wasn't in any way ground-breaking – they were first discovered in the 1970s – but we could now begin a raft of detailed studies in the hope of seeing some evidence of them misbehaving, at least as far as our accepted theory of particle physics is concerned.

D mesons only live for around a half a trillionth of a second, so they don't hang around in the wider world. They're created at the LHC when the enormous kinetic energy of two colliding protons is converted into new matter. Accompanying them is a veritable cornucopia of other particles that come flying out from the impact point like the glowing embers of a firework. Among the hundreds of different types are familiar ones like protons, neutrons, and electrons, but also ones with strange and exotic names: pions, kaons, lambdas, deltas, eta primes, rhos, sigmas, psis, phis, upsilons, xis, omegas. A typical collision looks like someone has stuck a stick of dynamite in a can of Greek alphabet soup.

What are all these particles and where do they come from? The answer to this question is deeply tied up with our search for the ultimate recipe for apple pie. It turns out that the protons and neutrons of which we are made are just two members of a much larger family of related particles that gradually began to appear in experiments from the 1930s onwards. Their arrival on the scene was unwelcome, at first at least, and caused no end of confusion for many years. But slowly and surely a pattern began to emerge that seemed to hint at a more fundamental structure. The discovery of what lay beneath would open the way to a far deeper under-standing of the nature of matter and would unlock the ultimate origin of the protons that make up our universe.

WHO ORDERED THAT?

Just around the corner from my office at the Cavendish is a corridor lined with wooden cabinets crammed full of what you might be forgiven for thinking is the kind of junk you'd find in your grand-dad's shed. In fact, if particle physics had a hall of fame, this would be it. Among the historical curios are the cathode ray tube that J.

J. Thomson used to discover the electron, Chadwick's battered brass tube that revealed the neutron, and at the far end a large bulb from the particle accelerator that first smashed the atomic nucleus to bits. Easily missed amid the experimental bric-a-brac is an unassuming brass and glass contraption that revolutionized particle physics: the first cloud chamber.

As the name suggests, the cloud chamber was originally designed to create artificial clouds in the lab after its inventor, the Scottish physicist Charles Wilson, fell in love with the dramatic atmospheric effects he saw while working on the summit of Ben Nevis in the Scottish Highlands. To test the idea that clouds form when water vapour attaches to airborne grains of dust, he built a water-vapour-filled chamber with as little contamination as possible, expecting that without any dust particles to act as seeds no clouds would be able to form. However, on examining the chamber he was surprised to see delicate wisps of water droplets streaming in every direction, like the contrails left by a fleet of tiny passenger jets. (Although since it was 1895, the comparison wouldn't have occurred to him.)

Completely by accident, Wilson had invented the first instrument capable of making individual subatomic particles visible to the human eye. Each fleeting track was caused by a single charged particle zipping through the chamber and knocking electrons off gas molecules as it went, leaving a trail of positive and negatively charged ions in its wake. These ions attract water molecules out of the vapour, growing until they form trails of droplets big enough to see.

The cloud chamber was a literal revelation. For the first time, physicists had a window into the hidden world of atoms and particles, allowing them to watch and even photograph them going about their otherwise invisible business. Ernest Rutherford called it 'the most original and wonderful instrument in scientific history' and it became the primary tool of subatomic physics for the first half of the twentieth century, leading directly to three Nobel Prize-winning discoveries.

One unarguable master of cloud chamber photography was the American physicist Carl Anderson. Anderson spent much of the 1930s using cloud chambers to take photographs of cosmic rays – particles that rain down on the Earth from outer space. In 1932 he had rocked the physics world when he snapped a photograph of the first antiparticle, a positively charged mirror image of the electron known as the 'positron'.

The positron wasn't entirely unexpected – the British theoretical physicist Paul Dirac had predicted its existence three years earlier – but in 1936 Anderson and his colleague Seth Neddermeyer would discover another particle that really upset the apple cart. A year earlier they had decided that to get better images of cosmic rays they needed to get closer to the source of the action, so they loaded their cloud chamber onto a flatbed trailer bought at a used-car lot near their lab at Caltech in Pasadena and set off for the Colorado Rockies. They erected their equipment on the summit of Pikes Peak, a 4,300-metre-high pink granite mountain near Colorado Springs, camping out each night in a bunkhouse halfway down the mountain. After months of long days and nights working at high altitude, they returned to Pasadena to develop their photographs and analyse the results. Examining the beautiful cloud chamber traces, each showing dozens of particle tracks curving elegantly in their powerful magnetic field, they discovered a particle unlike anything they'd seen before.

They convinced themselves that these new particles were neither featherweight electrons nor the relatively bulky protons. In fact, their rough measurements suggested that they had a mass somewhere between the two – about two hundred times heavier than an electron or about one-tenth the mass of a proton. Given its in-between mass, Anderson and Neddermeyer coined the term 'mesotron' – *mesos* in Greek means 'middle' – but today we know it as the muon.

The muon didn't seem to be a constituent of an atom – it only seemed to be found in cosmic rays – so what was it for? Well, at first at least, it looked to be a good match for a particle predicted by the Japanese theoretician Hideki Yukawa, who had been pondering the force that keeps protons and neutrons held together inside the atomic nucleus. Since protons are all positively charged, they should exert a stupendously large repulsive force on one another when squashed into a space as cramped as the nucleus of an atom. The only way nuclei could be holding together was if a much stronger attractive force between their constituents over-whelmed the electrical repulsion. The puzzling thing about this 'strong nuclear force', as it became known, was that it didn't seem to have any effect until two protons or neutrons were almost touching each other. At distances longer than about a thousandth of a trillionth of a metre the force seemed to disappear entirely.

How to explain the peculiar short range of the strong nuclear force? Yukawa's bright idea was that the force was communicated between protons and neutrons by the exchange of a new type of 'heavy particle', as Yukawa dubbed it. The fact that the proposed particle was heavy was key. The particle's large mass meant that it could only travel a very short distance,* severely limiting the range of the strong nuclear force. Based on measurements of how protons, neutrons, and nuclei bounce off one another, Yukawa calcu-lated that his particle needed to have a mass of 100 MeV; for reference, the electron's mass is 0.5 MeV, compared to 938 MeV for the proton.

At first it looked as though Anderson and Neddermeyer had bagged Yukawa's heavy particle – its mass appeared to be almost precisely in line with Yukawa's prediction. A surge of excitement swept through the physics community. At last the mysterious nature of the strong nuclear force might be within reach. However, doubts soon began to grow. For one thing, the particle they'd discovered

---

* The reason for this has to do with Heisenberg's uncertainty principle in quantum mechanics, which is a bit of a diversion for the purposes of our story right now. We'll come back to it.

seemed to be able to penetrate much farther through slabs of metal than you'd expect of the particle of the strong nuclear force, which should interact enthusiastically with atomic nuclei and get stopped much more abruptly. What's more, Anderson and Neddermeyer's particle lived far longer before decaying than Yukawa had predicted.

It would take more than a decade to clear up the confusion. In 1947 a group led by Cecil Powell at the University of Bristol used a completely different technique – based on exposing photographic plates to cosmic rays – to discover a new charged particle, which they called the 'π meson' but today is usually shortened to just 'pion'. Here at last was Yukawa's predicted carrier of the strong nuclear force! In fact, there were three types of pion, a positive and a negative one, along with an electrically neutral version discovered a few years later. Yukawa was quickly rewarded with a Nobel Prize for his audacious prediction, and Powell for the experimental discovery.

An enlarged picture of the make-up of matter was emerging. Electrons orbit atomic nuclei made of protons and neutrons, which are bound together in their nuclear prison by the frenetic exchange of three types of pion. Rather pleasingly, this means our apple pie is in part made from pions. Hanging around awkwardly on its own was Anderson and Neddermeyer's muon, which seemed to look a lot like a heavy and unstable version of the electron but without serving any useful function as far as anyone could tell. The physicist Isidor Rabi famously captured the confusion caused by the muon with the pithy phrase 'Who ordered that?' as if it were a pizza delivery that had turned up unexpectedly.

The appearance of the pions kicked off a flood of new discoveries. In the same year, the Manchester duo of George Rochester and Clifford Butler spotted strange pairs of forking tracks in their cloud chamber that seemed to be produced by the decay of a new particle about a thousand times heavier than the electron. Originally dubbed 'V particles' due to their distinctive V-shaped decays, they are now known as 'kaons'. Before long, physicists were faced with a plethora of proliferating particles: some lighter than protons and neutrons, others heavier.

What were all these new particles for? No one had the foggiest idea. Such was the confusion that one physicist quipped, 'The finder of a new elementary particle used to be rewarded by a Nobel Prize, but such a discovery now ought to be punished by a ten-thousand-dollar fine.' Physics appeared to be in danger of morphing from an elegant subject with only a small number of ingredients governed by a few simple, unifying principles into something more akin to zoology, with a baffling variety of different species jostling for space in an ever-growing particle zoo. Physicists, who often wear their inability to remember anything so banal as a list of facts, dates, or names as a badge of honour, were horrified. Enrico Fermi famously harrumphed, 'If I could remember the names of all these particles, I would have been a botanist!'

Amid the chaos, physicists did their best to impose some kind of order. There were a few clues to follow. First off, with the notable exception of the muon, all these new particles experienced the mighty pull of the strong nuclear force. To distinguish them from particles that didn't – like the electron, muon, or the photon – physicists dubbed this new family of strongly interacting particles 'hadrons'. The hadrons could be broadly divided into two further categories: those with masses in between the electron and the proton became known as 'mesons', while those heavier than the proton were classed as 'baryons'.

More could be gleaned by sorting the hadrons according to their essential properties or quantum numbers. One example that we've already come across is electric charge; the proton has an electric charge of +1, while Rochester and Butler's kaon has a charge of 0. Another extremely important one is a particle's angular momentum, or spin. If momentum is the amount of oomph carried by a particle due to its motion in a straight line, then angular momentum is the amount of oomph due to its rotation. Quantum mechanical spin comes in discrete lumps of size ½ and can only take values in the sequence 0, ½, 1, ³⁄₂, 2, ⁵⁄₂, and so on. A great deal of effort was spent figuring out the spins of the zoo of particles that continued to appear in experiments. At first the mesons all seemed to have spin 0 and the baryons spin ½, but before long mesons with spin 1 and baryons with spin ³⁄₂ were also discovered.

By the 1950s, physicists were no longer content to wait for particles to arrive in their cloud chambers from outer space. Now the pendulum swung towards huge accelerators, which could produce exotic particles on demand by firing protons or electrons into suitable targets, converting their kinetic energy into new particles in the process. In 1953, a hulking ring-shaped particle accelerator known as the Cosmotron was inaugurated at Brookhaven National Laboratory on Long Island, New York. The first accelerator to break the billion-volt barrier, the Cosmotron used a series of powerful magnets to guide beams of protons around in a circle so that they could be accelerated repeatedly each time they orbited the ring, reaching energies high enough to create the full panoply of particles that had previously only been seen in cosmic rays.

One of the Cosmotron's achievements was helping to pin down yet another property that some particles appeared to possess. Certain members of the particle zoo lived far longer before decaying than theorists naively expected, and what's more these particles always seemed to be produced in pairs. In 1953 the theoretical physicists Kazuhiko Nishijima and Murray Gell-Mann independently proposed that the reason these particles lived an unusually long time was because they carried a new quantum property, which given their strange behaviour they called simply 'strangeness', a term that has survived to this day. The Cosmotron could get protons to such high energies that it could recreate all the strange particles discovered so far in cosmic rays, along with a new strange meson that hadn't been seen before.

The Cosmotron was aided by the arrival of a brand-new type of detector that allowed physicists to record the cascades of decaying particles in unprecedented detail. These bubble chambers were descendants of the cloud chamber, but instead of gas they were filled with supercooled liquids – usually liquid hydrogen, freon, or propane. The liquid was held just below the boiling point until the physicists were ready to fire a beam of particles into the chamber, at which point the pressure was suddenly reduced, causing little bubbles of gas to erupt along the paths taken by electrically charged particles. At the same instant a flash of light was sent through the chamber, illuminating the beautiful trails of bubbles so that they could be captured by cameras peering in through portholes around the edge of the chamber.

The winning combination of record-breaking energy and a shiny new bubble chamber allowed the Cosmotron to steal a march on its competitors, but its success triggered a particle accelerator arms race. Ever larger and more powerful machines were soon being built around the world, many of them with excitingly futuristic names. Across the bay from San Francisco at Berkeley, the place where the first circular particle accelerator had been invented at the start of the 1930s, the Bevatron smashed the Cosmotron's record, reaching a beam energy of 6.2 gigaelectron volts (GeV)* and discovering the antiproton in 1955. Not to be outdone by its capitalist foe, the Soviet Union soon had its own superbly named Synchrophasotron in Dubna near Moscow, whose peak energy of 10 GeV left the American machines eating its dust. Europe briefly took the lead when it fired up its 28 GeV Proton Synchrotron at CERN in 1959 until it was knocked off the top spot by the United States' Alternating Gradient Synchrotron (AGS), built at Brookhaven in 1961.

The race for ever higher energy brought a flood of new particles, transforming these giant accelerator complexes into the boomtowns of particle physics, packed full of ambitious researchers hoping to sift some shiny new nugget from the subnuclear detritus. The particle zoo continued to grow apace, and yet as it did, what at

---

* A gigaelectron volt is a billion electron volts, the kinetic energy of an electron after it's been accelerated through a billion volts.

first had seemed to be unconnected fragments slowly began to come together to reveal hints of an underlying order. That said, big chunks were still missing, and the relationships between the pieces were clouded by the messiness of the experimental data. It would take a mind of extraordinary vision and clarity to see through the fog and discern the jewel-like symmetry beneath. Luckily, such a mind turned up in the form of Murray Gell-Mann.

ESCAPE FROM THE ZOO

Murray Gell-Mann grew up in Manhattan in the 1930s and 1940s, the son of Jewish immigrant parents from the Austro-Hungarian Empire. His older brother Ben taught him to read at the age of three using a Sunshine cracker box and introduced him to a love of bird- and mammal-watching, botany, and insect collecting. As kids, Murray and Ben would wander all over New York City in search of the few surviving fragments of unspoiled nature where they might spy an interesting animal or plant. Murray's orderly mind delighted in the way you could arrange all of the different living things they came across into species, connected together on the evolutionary tree.

In 1960 Gell-Mann, already one of the world's most respected theorists, had a flash of insight that would eventually unravel the mystery of the particle zoo. Like a zoologist sorting different species into genera and families, he began by arranging the known hadrons into their own broad groupings, the spin 0 mesons and the spin ½ baryons, and then searching for deeper connections between the individual members. Protons and neutrons seemed to make a neat pair, with almost the same mass but different electric charges, and since both had spin ½ they clearly belonged to the baryons. Then there were the pions, which came in positive, negative, and neutral varieties, and then two strange kaons, positive and negative, all of which were spin 0 mesons.

As he played his game of particle categorization, Gell-Mann became convinced that a deep symmetry was lurking just below the surface. In search of a structure that might describe the patterns he was seeing, he turned to what had been until recently a relatively neglected area of mathematics known as 'group theory'.

One of group theory's many applications is its use in describing symmetries. Simply put, a symmetry exists when you can do something to a system that leaves it unchanged. Take for example an ordinary cube. Since a cube has a high degree of symmetry, there are many ways of rotating it so that it ends up looking the same as it did before. These rotations form what is known as a group, which is just the collection of ways of rotating a cube while leaving it looking the same.

As he puzzled over the hadrons, Gell-Mann thought he spied the imprint of a more abstract mathematical group known as SU(3). Unfortunately, there's no easy-to-imagine way of describing SU(3) without resorting to mathematics, but the important point is that Gell-Mann realized you could use the symmetries of the SU(3) group to arrange the hadrons on a set of grids according to their spins, electric charges, and strangeness, producing sets of hexagons with a particle at each corner and two in the centre.

By ordering the hadrons in this way Gell-Mann did for them what Dmitri Mendeleev had done for the chemical elements a century earlier. Just as Mendeleev had predicted the existence of new elements from gaps in his periodic table, Gell-Mann was able to predict the existence of new hadrons. The symmetries of the SU(3) group required there to be eight spin 0 mesons and eight spin ½ baryons, but so far only seven spin 0 mesons had been discovered.

When Gell-Mann published his theory in 1961, he found that another physicist, Yuval Ne'eman of Imperial College in London, had come up with the same idea at almost exactly the same time. However, Ne'eman was a relative unknown, having only recently left the Israeli military to enter the world of physics, while

Gell-Mann was already a highly respected figure and an able communicator to boot, which ensured that his version of the theory reached a much wider audience.

Erudite as well as brainy, and not shy about showing it off, Gell-Mann turned to ancient Buddhist teachings to find a name for his theory, which he dubbed the 'Eightfold Way' after the path that liberates those who follow it from the endless cycle of death and rebirth. When just months later the missing eighth meson, named the 'eta meson', was found by the team at Berkeley, physicists began to believe that Gell-Mann might have found the way to hadronic nirvana.

However, the clincher really came with the discovery of a clutch of new, even heavier particles. As well as predicting octets of spin 0 mesons and spin ½ baryons, the Eightfold Way required that there should be ten baryons with spin ³⁄₂. When these spin ³⁄₂ particles were arranged on the same grid of electric charge versus strangeness, they traced out the shape of a pyramid. At the time that Gell-Mann and Ne'eman published their theories, only four such particles were known, the so-called Delta baryons with 0 strangeness, which presumably made up the base of the pyramid. Then, in July 1962 physicists thronged to a major conference hosted at CERN where particle hunters announced solid evidence for three new Sigma-star baryons with strangeness −1 and a pair of Xi-star baryons with strangeness −2.

Gell-Mann and Ne'eman knew immediately that these five new particles must form the next two layers of the pyramid. After the discoveries had been announced, Gell-Mann leaped to his feet to predict the existence of the tenth and final missing particle, the capstone of the pyramid with strangeness −3, which he named the 'Omega' after the last letter of the Greek alphabet. Ne'eman, who had also raised his hand to speak but was sitting farther back in the hall, could only watch on gloomily as Gell-Mann made the very prediction he had been about to propose.

Later, over lunch with two young experimenters from Brookhaven,

Nicholas Samios and Jack Leitner, Gell-Mann grabbed a napkin and sketched out how the Omega might be found by looking for its likely decay products. The pair took the napkin back to Brookhaven and used it to persuade the lab's director to give them time on the AGS, the most powerful particle accelerator in the world. After more than a year spent getting the accelerator and bubble chamber into working order, the team began collecting data just before Christmas, working feverishly around the clock into the new year. Poring over tens of thousands of bubble chamber photographs, each criss-crossed by numerous particle tracks, Samios spotted a single image with multiple strange particles all pointing back to a common point of origin, the smoking gun of the Omega.

The discovery of the Omega sealed the deal for the Eightfold Way. By 1964 there was a powerful sense that another great revolution in our understanding of the subatomic world was underway. At long last the particle zoo was being tamed.

But what did it all mean? As we've seen, the patterns in Mendeleev's periodic table were the first clues that supposedly indivisible atoms actually have an internal structure, which ultimately determines each chemical element's unique properties. Could the Eightfold Way be hinting at something similar? Could all these hadrons, including the protons and neutrons that make up the chemical elements, be made of even smaller things?

Not necessarily. The most popular explanation of the existence of the hadrons at the time did away with the distinction between fundamental particles with no internal structure and composite particles made of smaller things. The American theoretical physicist Geoffrey Chew instead argued for what he called 'nuclear democracy', where no particle could be thought of as being any more fundamental than any other. According to Chew, each hadron was a mixture of all the others.

This fantastically counter-intuitive idea became known as the 'bootstrap model', as it involved the hadrons effectively pulling themselves into existence, like the nonsensical idiom of pulling yourself up by your own bootstraps. The great hope of the bootstrap theorists was that there might only be one possible set of hadrons that could pull themselves into existence, in which case you'd have a fantastically economical theory explaining all the known particles without any external inputs. Perhaps the Eightfold Way was a consequence of the deeper truth provided by the bootstrap model, which many hoped would soon come into view.

However, the bootstrap model wasn't the only show in town. For the past few years Murray Gell-Mann had been playing around with the idea that the symmetries he'd spied in the hadrons could be explained if you thought of them as being made up of smaller bits. He had never taken the idea very far, partly because he thought it was incompatible with the more aesthetically pleasing bootstrap model and partly because he was busy solving other more pressing problems. Also, these smaller bits, whatever they might be, would need to have fractional electric charges of $\frac{1}{3}$ or $\frac{2}{3}$ that of the electron's, but so far, the charge of every particle seen in nature was a whole number.

In March 1963, Gell-Mann was having lunch with some colleagues at Columbia University in New York when he got chatting with the physicist Robert Serber, who had also been thinking about subhadronic building blocks. Gell-Mann was dismissive when Serber asked him what he thought about the idea over lunch, but later that evening their conversation got Gell-Mann thinking. What if these fractionally charged nuggets were forever locked up inside hadrons, never able to escape into the external world? If that were true, the cherished principle of nuclear democracy could be preserved and the bootstrap model would still be viable.

Gell-Mann, who had a talent for conjuring memorable monikers, dubbed these undetectable little particles 'qworks' – a kind of nonsense word in the style of Lewis Carroll. Months later, while perusing the notoriously incomprehensible novel *Finnegans Wake* by James Joyce, his eyes alighted on the phrase 'Three quarks for Muster Mark!' amid Joyce's gibberish. Gell-Mann immediately realized he had found a perfect opportunity to give his little building blocks a literary heritage and, more importantly, taking their names from such an obtuse work would only further impress upon his colleagues just how well read and clever he really was. And so 'qworks' became 'quarks'.

According to Gell-Mann, the symmetries in the hadrons could be explained if there were three such quarks, which he called 'up', 'down', and 'strange'. The up quark had a charge of $+\frac{2}{3}$, while the down and strange quarks both had charges of $-\frac{1}{3}$. By combining these three particles (and their antiversions) you could explain the properties of all the known hadrons. Mesons like the pion or the kaon were a pairing of a quark and an antiquark, while the baryons were a threesome of quarks. Most importantly for our purposes, the proton could be thought of as being made of two up quarks and a down quark, while a neutron was built from two downs and an up.

Meanwhile, several thousand kilometres away at CERN near Geneva, a young Russian postdoc and former PhD student of Gell-Mann's named George Zweig was thinking along very similar lines. Completely independently, Zweig had realized that the symmetries of the Eightfold Way could be explained if there existed three individual building blocks with electric charges $+\frac{2}{3}$, $-\frac{1}{3}$, and $-\frac{1}{3}$, which he called 'aces'.

However, while the two ideas were identical when it came to accounting for the symmetries in the hadrons, Zweig and Gell-Mann had very different takes on what it all actually meant. Gell-Mann was happy to think of the quarks as mathematical conveniences rather than real physical entities. The really fundamental ingredients of hadrons, as far as he was concerned, were the mathematical symmetries that they appeared to obey. Quarks were just a convenient way of keeping track of these fundamental symmetries but would probably never be observable in the real world.

For Zweig, on the other hand, quarks (or aces) could be just as real as protons, neutrons, and electrons. Unfortunately for the young physicist, such ideas were wildly unfashionable at a time when the weird but elegant bootstrap model was in vogue. To argue that the hadrons were made of smaller things seemed simpleminded, even childish. Gell-Mann himself teasingly referred to Zweig's aces as 'the concrete block model'. As a result, while Gell-Mann had no trouble getting his quark theory published in a respected journal, Zweig faced such a barrage of criticism from his referees that his paper never saw the light of day, except as a lowly CERN preprint, an article put out by the lab itself rather than being published in a prestigious journal.

However, while some theorists were sniffy about the idea of quarks, there were plenty of experimentalists for whom the prospect of discovering a new layer of reality was too good to pass up. Physicists began poring over tens of thousands of old bubble chamber photographs in search of fractionally charged particles that they might have missed. New particle beam experiments were hastily prepared at CERN and Brookhaven in the hope of spotting a quark being knocked free of one of its hadrons. Even some die-hard cosmic ray physicists got in on the act, searching for quarks amid the showers of particles that rain down from the heavens.

But quarks were nowhere to be found. By 1966 twenty experiments had searched for them and come up empty-handed. Speaking at the Royal Society in London that year, Gell-Mann himself declared that 'we must face the likelihood that quarks are not real'.

Help would come from an unexpected source. At Stanford University in Northern California, the finishing touches were being made to the world's largest and most expensive particle accelerator. Stretching 3.2 kilometres in a straight line through the rolling parkland of the Stanford campus and passing directly under Interstate 280, the Stanford Linear Accelerator was effectively an enormous particle cannon, capable of accelerating electrons to a whopping 20 GeV. Its enormous scale and $100 million price tag had earned it the nickname 'the Monster', and getting it built had taken more than a decade of planning, design, and construction, not to mention navigating the project through the US Congress.

At a time when most physicists were focused on the exciting new discoveries emerging from the high-energy proton accelerators at CERN and Brookhaven, the Monster was a bit of an odd beast. Unlike its circular cousins, which worked by steering beams of protons around a ring and accelerating them each time they completed an orbit, the Monster fired electrons down a dead-straight 3.2-kilometre tube,* accelerating them all the way until they slammed headfirst into a target at the far end. The aftermaths of these collisions were then recorded by towering spectrometers, which measured the energies and directions of the scattered electrons.

In effect, the Monster was a colossal microscope capable of zooming right in on the proton to study its size and shape in unparalleled detail. The higher the energy of an electron beam, the shorter the distances it can probe, resolving ever finer details. The reason that higher-energy particles allow you to explore shorter distances is down to the quantum mechanical phenomenon of wave–particle duality – specifically that a particle like an electron can be caught behaving like a wave, if you set up an experiment in the right way. The wavelength of an electron, or indeed any particle, depends inversely on the momentum of the particle; in other words, the faster the particle is moving, the shorter its wavelength.

When it fired up in 1966, the Stanford Monster could accelerate electrons to 99.99999997 per cent of the speed of light, giving them a wavelength of about 6×10^{-17} metres (sixty-millionths of a trillionth of a metre). Experiments had shown that protons and neutrons were about 1×10^{-15} metres across, and so in principle the Monster's beam could resolve objects far, far smaller than these most basic building blocks of atoms.

In the mid-1960s, theorists imagined the proton as a fuzzy, insubstantial sphere with no internal structure. As a result, when they

---

* At the time it was claimed to be the world's straightest object.

fired their super-high-energy electron beam at the proton, the team working on the Monster expected most of the electrons to zip right through almost unimpeded. Remind you of anything?

Back at the start of the twentieth century, physicists had pictured the atom as a similarly insubstantial pudding-like object, which is why Ernest Rutherford had been so thunderstruck when alpha particles came bouncing straight back off gold atoms. That famous result had completely changed our understanding of atoms, eventually leading the boisterous New Zealander to conclude that the atom has a tiny nucleus at its heart.

Something eerily similar was about to happen at Stanford: their giant accelerator was really just Rutherford's gold foil experiment writ large, albeit on a scale totally unimaginable in 1908. Sixty years after the discovery of the nucleus, physicists were still using Rutherford's tried and tested technique of firing particles at a target and seeing how they bounced off.

Stanford even had its own version of Rutherford in the fearsome figure of Richard Taylor, a towering presence whose angry, booming voice could often be heard echoing along the corridors. After the first set of electron scattering experiments ended in 1966, Taylor took charge of a joint Stanford–MIT team who began to probe ever deeper into the proton. In 1967, they got the first hints that something strange was going on. Electrons seemed to be losing far more energy as they passed through the proton than expected.

At first the effect was dismissed as noise, but by early 1968 the team had convinced themselves that what they were seeing was real. Just like Rutherford's alpha particles, the electrons were being scattered through far larger angles than you would expect if the proton really was just a diffuse sphere of electric charge. There seemed to be only one explanation – the electrons were bouncing off unimaginably tiny objects *inside* the proton.

Against anyone's expectations, this giant accelerator had peered deep within the most basic building blocks of matter and glimpsed a brand-new layer of reality. Despite the popularity of fancy ideas like the bootstrap model, the old, tried, and tested atomic view of matter appeared to have won yet again. Protons, neutrons, and all the hadrons in the particle zoo really did seem to be made of even smaller particles.

However, the Stanford–MIT team had a fight on their hands persuading people that they had really seen quarks. Such was the hold of the bootstrap model that at first their electron scattering results aroused little interest. It would take years more experimental and theoretical work, not to mention the enthusiastic advocacy of physics' most charismatic communicator, Richard Feynman, to convince the world that the Monster really had seen the building blocks of the proton.

It was in 1973, after CERN's gargantuan bubble chamber named Gargamelle spotted neutrinos ricocheting off point-like objects inside the proton, that the evidence for quarks became overwhelming. Comparing Gargamelle's and the Monster's results, physicists were able to discern three such particles within the proton; what's more, these particles appeared to have fractional charges, just as Gell-Mann and Zweig had predicted. Despite Gell-Mann's scepticism about the reality of his own inventions, quarks had finally become real physical objects that physicists could begin to believe in.

Well, sort of. One great puzzle remained – no one had actually seen a quark. All the evidence for their existence came from bouncing particles off hadrons. No accelerator, no matter how powerful, had managed to break a single, solitary quark from its hadronic jail cell. Quarks seemed to be inexorably locked up inside.

The reason, it turns out, has to do with the force that binds quarks together inside hadrons. This force – known simply as the strong force – is the most potent attraction ever discovered. The strong *nuclear* force that holds protons and neutrons together inside atomic nuclei is a kind of echo of this far mightier force. To break the bonds of the strong force and liberate quarks from inside protons and neutrons requires temperatures far hotter than the hottest star, temperatures of trillions of degrees.

Such temperatures have not been seen in the universe since a millionth of a second after the big bang. It was during this first microsecond of cosmic time that the protons and neutrons from which we are made came into existence. To get at the ultimate origins of matter we'll need to find a way to probe the physics of this trillion-degree universe. Incredibly, such temperatures are now routinely recreated here on Earth, just a few kilometres from the bustling heart of New York City.

TRILLION-DEGREE SOUP

For a country that prides itself on being a freewheeling, beacon-of-liberty, keep-government-out-of-my-business, who-is-the-federal-government-to-tell-me-I-can't-own-a-surface-to-air-missile? sort of place, the United States can be surprisingly officious. Ahead of my visit to Brookhaven National Laboratory I was required to fill out a multipage online application form, followed by a fairly lengthy back and forth with their (unfailingly helpful) administrative staff about the purpose of my visit. Crucially, I was told that on entering the United States I should take great care to get the correct stamp at immigration – get it wrong and I wouldn't be allowed access to the site. So ensued a meandering conversation with two slightly bemused-looking border officials about what exactly I was doing in their country, during which I tried to explain that I just wanted to visit some government labs and chat to some scientists in a completely innocuous, non-espionagey kind of a way, while taking great pains to avoid using the word 'nuclear'. By comparison, it used to be possible to get onto the CERN site by waving your Tesco Clubcard at an uninterested-looking security guard.*

So it was with some trepidation that I presented myself at the

---

* Before you try to break into CERN, I should add that things have been tightened up a bit since.

security hut on the wooded road running into the Brookhaven site, brandishing my passport complete with a worryingly faint immigration stamp. The woman on the desk eyed it suspiciously. 'I think they were running out of ink,' I said, smiling weakly. After some tutting and a bit of tapping on a computer, to my relief her face brightened, and she handed back my passport. 'Welcome to Brookhaven.'

Brookhaven National Laboratory has a long and illustrious history when it comes to particle physics. Founded in 1947 on the site of an old US Army training camp, its first major facility was an experimental nuclear reactor, followed by the billion-volt-barrier-busting Cosmotron accelerator in 1953, which played a leading role in exploring the particle zoo. Then in 1960 came the Alternating Gradient Synchrotron (AGS), which ruled the roost as the world's highest energy accelerator for the best part of a decade.

Among the AGS's many achievements was a major discovery that sent particle physicists into a frenzy of excitement in 1974. The November Revolution, as it's known in the field, began when a team led by Samuel Ting at Brookhaven discovered a striking new peak in their data at an energy of around 3.1 GeV, or just over three times the mass of the proton. Meanwhile, 4,000 kilometres away in California, Burton Richter's group working on the Stanford Monster were staring astounded at exactly the same spike. Both groups announced their discoveries on 11 November. The peak proved to be evidence of a hadron made from a brand-new, never-before-seen type of quark – the 'charm quark' – a heavier cousin of the positively charged up quark found inside protons and neutrons.[*]

The AGS's discovery removed all remaining doubts about the existence of quarks and did much to lay the foundations of our current theory of particle physics. Today, this venerable accelerator

---

[*] Today, we know of six quarks in total. The up, charm, and top quark with electric charges of $+\frac{2}{3}$ and the down, strange, and bottom quarks with electric charges of $-\frac{1}{3}$.

is still in operation as the feeder for an even larger and more powerful atom smasher – the Relativistic Heavy Ion Collider, or RHIC for short. It was this machine that I had come to see.

To understand what the scientists at RHIC are up to, we need to delve a little further into the physics of the quarks that make up protons and neutrons. At the same time that the reality of quarks was becoming accepted in the early 1970s, physicists were trying to understand the mysterious strong force that keeps them locked up inside hadrons.

By 1973 a candidate theory had emerged based on the very same SU(3) symmetry group that Gell-Mann and Ne'eman had used to categorize the hadrons in the Eightfold Way. However, this time the symmetry described the strong force itself.

Just as protons and electrons attract each other through the electromagnetic force thanks to their opposite electric charges, quarks attract each other because they carry the equivalent charge of the strong force. But whereas there is only one kind of electric charge, which can be either positive or negative, SU(3) symmetry dictates that the strong force should have three different types of charge, each with its own positive and negative version. Again demonstrating his uncanny talent for picking terms that stick, Gell-Mann called these three strong charges 'colours'. Not to be confused with actual colour like the colour of my sweater (orange, in case you were wondering), the colour of a quark is just a word for the charge that determines how it feels the strong force. Originally Gell-Mann patriotically suggested that these three colours should be called red, white, and blue, but today physicists usually plump for the more neutral red, green, and blue.

If quarks come in red, green, and blue varieties, antiquarks come in anti-red, anti-green, and anti-blue, and, just as with electric charge, like colours repel while opposites attract. So two red quarks will repel each other while a green quark and an anti-green antiquark will want to get together. One of the things that makes the strong force more complicated than the electromagnetic force is that the three different colours also attract one another, so a red up quark, a green up quark, and blue down quark will draw one another together to form a proton. Hadrons (particles made from quarks) are always colourless overall; either a colour paired with its anti-colour in a meson, or all three colours mixed together in a baryon. Thanks to all this colour business, the theory has a very cool-sounding name: 'quantum chromodynamics' (QCD), the quantum theory of colour.

As well as dictating that quarks come in three colours, QCD tells us that the strong force is transmitted by particles called 'gluons', literally because they glue quarks together. At first glance, gluons look a lot like photons, the force carriers of electromagnetism. Like photons they have zero mass and a spin of 1. However, the particular requirements of the SU(3) symmetry group mean that while there is only one type of photon, there are *eight* different types of gluon. And, crucially, while the photon carries no electric charge, gluons are coloured. It's this final fact that explains why, even to this day, no one has ever seen a quark flying solo.

Here's why. Photons only interact directly with electrically charged particles like protons and electrons. Since photons are electrically neutral, this means that if you fire two photons at each other they'll (almost) always just zip past each other without so much as a gentle handshake. They pass like ships in the night.

Things are different for the gluons. Each gluon carries a combination of colour and anti-colour, and since gluons are attracted to coloured particles, they will actually interact *with each other*. This means that the strong force between two quarks is completely different from the electromagnetic force between, say, a proton and an electron.

We're almost at the point of understanding why no one has ever seen a naked quark, as it were, so bear with me. Imagine an electron and a proton sitting a little distance apart, like they might in a hydrogen atom. One way to conceptualize the electromagnetic force between them is to imagine the proton and electron both firing off photons in every direction,* a bit like one of those spherical lights you sometimes still see at 1980s-style discos. Since the proton and electron are close together, a large number of the photons emitted by the electron will be attracted towards the proton and absorbed, and vice versa. It's this exchange of photons that creates the attractive force between the two charged particles.

Now imagine that we grab hold of the electron and proton and start to pull them apart. As the distance between them increases, fewer and fewer of the emitted photons will be absorbed by the opposing partner, and the attractive force between the electron and proton gets weaker and weaker. At first you have to work hard against the attractive force, but as you separate the two particles, it gets easier and easier until eventually you're left with a free electron and a free proton.

Now let's consider the equivalent situation for two quarks. Instead of photons, the two quarks now fire off gluons in every direction. Gluons that get shot out in the direction of the other quark get attracted to it and absorbed, creating an attractive force just like with the proton and electron. However, this is where the fact that gluons carry colour starts to change things. The exchange of gluons creates an excess of colour in the region between the two quarks. You can picture the gluons flowing back and forth between the two quarks as a tube of red, green, and blue, with a quark on either end. This colourful tube attracts other nearby gluons, drawing them into the gap and making the tube even denser and more colourful. Eventually there is so much colour in the tube that *all* of

---

* There is a subtlety here – the photons fired out by the particles in this example aren't real, observable photons like the ones produced by a lightbulb. Instead, they're what we call 'virtual' particles. Virtual particles are completely undetectable and are only really a crutch for thinking about how forces are transmitted between particles. To be honest, I don't find the concept of virtual particles particularly helpful – a far better explanation involves physical entities called 'quantum fields', which we'll come to soon enough – but they *are* useful for the purposes of this analogy.

the gluons emitted by both quarks get sucked in – forming a mighty multicoloured bond between the two quarks.

Now let's say we decide to try to separate the two quarks. We grab hold and start to pull. It's bloody hard work, but gradually the quarks start to creep apart. However, since all the gluons are still concentrated in a tube between the two quarks, the force we're fighting against doesn't get any weaker. Instead, the gluon tube stretches like an elastic band, and just like an elastic band, as we stretch it, more and more energy gets stored in the tension of the tube. Now – and this is the fun part – once the amount of energy stored in the tension of the gluon tube is equal to the mass of a new quark–antiquark pair, the tube catastrophically snaps, but instead of ending up with two free quarks, a new quark and anti-quark are created from the energy stored in the stretched gluon tube, each attaching itself to one of the broken ends. What we've got are two pairs of quarks, each of them still locked firmly together.

This is the reason we have never seen a bare quark. Try to pull a quark out of a hadron and, like magicians pulling handkerchiefs out of their sleeve, you instead end up with an ever-growing chain of hadrons, which gets longer and longer the harder you pull. When we smash protons together at the LHC, instead of knocking quarks out, we end up with great jets containing dozens of hadrons, all created from the energy of the initial kick that sent the original quarks flying apart.

From this kind of argument, it seemed that quarks were doomed to remain trapped inside hadrons forever. But in 1973 the theorists David Gross, Frank Wilczek, and David Politzer made a stunning discovery about the nature of the strong force. They calculated that as you collide hadrons at ever higher energies, the vicelike grip of the strong force should start to weaken. This implied that at suffi-ciently high energies, the strong force becomes so weak that hadrons effectively melt, turning into a superheated gas of free quarks and gluons.

This superheated stuff is known as a 'quark–gluon plasma', a stupendously hot and dense state of matter where quarks and gluons are finally free to zip about outside the confines of individual hadrons. To create one requires temperatures and densities far beyond anything that had ever been achieved in the lab in the mid-1970s. In fact, there was only one time in the universe's history when conditions had been extreme enough to create a quark–gluon plasma – that crucial first millionth of a second after the big bang.

Back then, the universe was so hot and dense that no hadrons could form; the entirety of space would have been filled with this seething mass of quarks and gluons. However, as the universe expanded it cooled, and after about a microsecond the temperature dropped low enough for quarks and gluons to fuse together to form the first protons and neutrons. That means that if physicists wanted to understand the ultimate origins of matter, they would need to find a way to study a quark–gluon plasma in the lab.

Enter RHIC, a 4-kilometre-circumference collider buried in a shallow tunnel cut through the soft sandy soil of Long Island. The principle of RHIC is similar to any other collider: two beams of particles are fired around the roughly hexagonal ring, one going clockwise, the other anticlockwise, kept on course by powerful electromagnets. On each orbit of the ring, high voltage electric fields give a kick to the particles as they pass by, gradually increasing their energy. Once the particles have reached the desired energy, the paths of the two beams are adjusted using magnets until they collide head-on inside large detectors, whose job is to record the subatomic debris that comes flying out from the collisions.

What makes RHIC different from other colliders is the projectiles it uses. As the name – Relativistic Heavy Ion Collider – suggests, RHIC's primary goal is to collide ions* of heavy elements, including aluminium, copper, uranium, and, sexiest of all, gold. The nuclei of these elements contain hundreds of protons and neutrons, and

* In this case an ion is an atom that has been stripped of some electrons, giving it an overall positive charge.

so when they collide enormous densities are created, potentially high enough to produce a quark–gluon plasma.

I had come to Brookhaven to meet Helen Caines and Zhangbu Xu, the two leaders – referred to as 'spokespeople' – of the STAR experiment,* one of two large detectors used to study the collisions produced by RHIC. We met over coffee in a large reception building close to the entrance to the Brookhaven site, a collection of office buildings and experimental halls spread across 21 square kilometres of land surrounded by dense woodland.

Sitting among the hubbub of Brookhaven staff getting their first vital caffeine injection of the day, Helen and Zhangbu talked me through the highs and lows of two decades studying the universe's most extreme state of matter. Helen cut her teeth as a PhD student at the University of Birmingham before crossing the Atlantic for her first research job in 1996. For someone interested in quark–gluon plasma, there was no better place to be in the late 1990s. RHIC was only a few years away from delivering its first collisions, and as a young researcher, Helen got in on the ground floor, joining the STAR Collaboration as soon as she arrived in the United States. At the time, her future co-spokesperson, Zhangbu, was working on his PhD at Yale, having originally studied physics in his native China. When data collecting began at RHIC, the two young physicists would be perfectly placed to lead the search for quark–gluon plasma.

However, before the experiments kicked off, the physicists at RHIC found themselves having to deal with unexpected headlines in the press courtesy of Hawaii resident Walter L. Wagner. Wagner was worried that the high-energy collisions at RHIC might end up destroying the world and obligingly provided a menu of doomsday scenarios to choose from. The collider might produce a tiny black hole that would gobble up the Earth or perhaps synthesize a new form of 'strange matter' that would convert the planet into a

---

* If you're into acronyms, STAR stands for Solenoidal Tracker at RHIC.

formless blob. Most exciting of all was the prospect of creating a bubble universe with different laws of physics, which would then expand at the speed of light, destroying not just our planet, but *the entire cosmos*.

The theoretical physicist Frank Wilczek jumped in to debunk Wagner's concerns, but that only seemed to fuel media interest, and eventually Brookhaven was forced to produce a lengthy report detailing why their new collider was unlikely to lead to the end of days.* Things calmed down after that, but it didn't stop Wagner filing twin lawsuits in New York and San Francisco in an attempt to halt the start of collisions. Fortunately, when the first gold nuclei smashed into one another at Brookhaven on 12 June 2000, the world carried on turning.

In the early days following the start of data collecting, some theorists were keen to claim that a quark–gluon plasma had already been created by RHIC based on the measurements taken by STAR and the three other detectors operating at the time. However, Helen, Zhangbu, and their experimental colleagues were far more cautious.

The great challenge of knowing whether you've made a quark–gluon plasma is that it's impossible to measure its properties directly. When two gold nuclei collide at RHIC the superheated blob of matter they form only exists for an instant. After a mere ten-trillionths of a trillionth of a second, this tiny fireball expands and cools down, transforming into an explosion of thousands of hadrons that tear through the detector at close to the speed of light.

These hadrons are all that STAR sees. It's only by studying their properties that you can infer whether a quark–gluon plasma formed. However, as time passed, the physicists at RHIC started to see telltale signs. First of all, they found that the thousands of hadrons

---

* The main reason why RHIC was unlikely to destroy the world was that cosmic rays with far higher energies than the collisions at RHIC have been bombarding the Earth, Moon, and other celestial bodies for billions of years. If producing world-destroying black holes, strange matter, or bubble universes was possible it would already have happened and we wouldn't be here.

seen by their detectors in each collision were flowing out from the impact point in a collective way, like the movement of a herd of wildebeest across a plain, strongly implying that they all originated from a single unified blob of matter. What's more, the number of jets produced by each collision was far lower than expected, almost as if the quarks were getting slowed down as they waded through a thick quark–gluon soup preventing them from converting their kinetic energy into jets of hadrons.

It took five years for the physicists at RHIC to be sure, but in 2005 they were ready to announce to the world that they had pulled it off – they had created a state of matter that hadn't existed in the universe since the big bang. They estimated that the quark–gluon plasma they'd produced had a temperature of around 2 trillion degrees, 130,000 times hotter than the centre of the Sun, and had a density of around a billion tonnes per cubic centimetre.

Most extraordinary of all, its bulk properties were completely different from what they had been expecting. Rather than a gas of free quarks and gluons, it behaved like a liquid, and not just any liquid, but a near-perfect fluid. This strange substance seemed to flow without any internal resistance or stickiness (or to put it more technically, it has almost zero viscosity). During the first microsecond the universe was filled, not by a fireball, but with a trillion-degree soup.

Our coffees drunk, we bid goodbye to Zhangbu, and Helen took me off to see STAR in the flesh. On the way we picked up her colleague and technical coordinator of the experiment, Lijuan Ruan. Like Zhangbu, Lijuan hails from China and arrived at Brookhaven as a young graduate student in 2002. Since then she's been deeply involved in all aspects of the experiment but particularly likes getting her hands dirty. Her delight and pride in the detector were evident: 'It's only getting hands on with the hardware that you really start to get a feel for how the whole thing works.'

The huge hall housing the STAR detector was on the other side of the campus so we jumped in Helen's car for a short drive, passing the aptly named Thomson Road and Rutherford Drive along the way. First stop was the control room, a dark bunker-like room with dozens of ancient-looking rear-projection computer screens used to monitor the performance of the experiment. Compared to the gleaming modernity of the LHC, the whole place had a distinctly well-worn feel. Unsurprising, perhaps, for an experiment about to enter its third decade.

From the control room we walked into a large hangar with a heavyset shield wall made from monolithic concrete blocks at one end. Much to my surprise there was no iris scan or radiation procedures to follow – the levels of radioactivity are well within safe limits as long as RHIC isn't running – and before I knew it, I found myself standing beneath the hulking mass of the STAR detector.

Weighing in at 1,200 tonnes and the size of a three-storey building, the detector makes quite an impression when you first come face-to-face with it. The bulk of the barrel-shaped detector is made up of a huge electromagnet used to bend the particles as they come flying out from the collision point, allowing physicists to measure their momenta. Nestling inside the magnet is the delicate STAR tracking system, which reconstructs the trajectories of the thousands of charged particles that are released as each tiny blob of quark–gluon plasma expands and cools. On the day I visited, STAR had been opened up, allowing me to see into the heart of the detector, complete with twinkling LED lights that made it look like something out of a science fiction movie.

As we stood on a raised gantry looking into the glowing heart of the detector, Helen and Lijuan told me about their plans for the next run of RHIC and the STAR experiment. Now that they're routinely able to create and study quark–gluon plasma, the team are closing in on a crucial moment in our universe's history, and one that's critical to our story. Around a microsecond after the big bang, the temperature of the universe dropped enough that the quark–gluon plasma transformed into the first protons and neutrons.

This is what physicists refer to as a 'phase transition', much like a liquid freezing to form solid ice. The plan for the next experimental run is to use RHIC to continually adjust the energy of the collisions, which roughly corresponds to varying the temperature of the quark–gluon plasma. The higher the energy of the colliding ions, the higher the temperature.

By slowly scanning through the collision energy, Helen and her colleagues hope to pinpoint the tipping point where a quark–gluon plasma 'freezes' to form hadrons. Figuring out how this process happened – effectively how protons and neutrons were cooked in the big bang – could have a profound influence on our understanding of how the first elements formed.

Having led the world in particle physics for the second half of the twentieth century, RHIC is now the United States' only remaining particle collider. For some years there were serious doubts about whether the research programme led by STAR and its friendly rival and neighbour on the ring, PHENIX,* would continue to be funded. During the 2000s, RHIC was the only show in town when it came to studying quark–gluon plasmas, but in 2010 CERN's Large Hadron Collider got in on the action with its own dedicated heavy ion experiment, ALICE.† In 2012, the far higher energy of the LHC allowed ALICE to smash RHIC's record for the highest ever recorded temperature, when lead ion collisions provided by the LHC produced a quark–gluon plasma with a temperature of more than 5.5 trillion degrees.

But while the LHC may dwarf RHIC in size and energy, there are still a few tricks its European rival cannot match. In particular, RHIC's ability to reduce its collision energy to lower values than the LHC means it's the only collider that can search for the critical

---

* Wondering what PHENIX stands for? Well it's Pioneering High Energy Nuclear Interaction eXperiment, apparently.
† ALICE stands for A Large Ion Collider Experiment – a rare example of an acronym for a particle physics experiment that actually works.

point when free quarks and gluons fuse to form hadrons. In the short term at least, the funding situation for America's last collider looks pretty rosy. With luck, it won't be long before Helen, Zhangbu, Lijuan, and their colleagues close in on the ultimate recipe for a proton.

CHAPTER 9

What Is a Particle, Really?

Our apple pie ingredient list has shrunk. A lot. We started with a whole cupboard full of ingredients – oxygen, carbon, hydrogen, sodium, nitrogen, phosphorous, calcium, chlorine, iron, and more besides – but are now left with just three: electrons, up quarks, and down quarks. That's cheating a little bit, because to bind these matter particles together to form atoms we also need the electromagnetic and strong forces. So add to that list photons and gluons. But still, that's a wonderfully economical list of basic ingredients, given that you can use them to make literally *anything*, apple pies included.

Quarks and electrons are particles, a term that I've admittedly used rather too casually so far, assuming that you've probably been picturing a little spherical thing, maybe a bit like a marble. As we've delved deeper into the structure of matter, that sort of mental image has served us pretty well; in many ways particles really do behave like little hard balls that stick together to create nuclei and atoms. The objects that come flying out from the collisions at the Large Hadron Collider travel through our detectors like microscopic*

---

* Another misused word. 'Micro' refers to objects a millionth of a metre in size; however, the proton is around 10^{-15} metres across, so really the correct term should be 'femtoscopic'.

bullets. When we create images of these collisions – usually only for publicity purposes these days; there are far too many to actually examine them all by eye – we draw out the path each particle took as if it really were a well-defined little nugget, as the word suggests.

This nuggetty picture of matter has a long pedigree; you can trace it back to John Dalton's atomic theory, and if you're trying to show off, all the way back to the ancient Greek philosophers Democritus and Leucippus, who were the first to argue that matter was made of indivisible, hard, particle-like things. However, the modern conception of a particle is a far cry from what Dalton or the ancient atomists imagined. The word 'particle' has become like an iceberg; its everyday meaning is just the visible bit above the waterline, while below the surface lurks a vast mass of properties, concepts, and half-understood phenomena that have built up over decades of experimenting and theorizing. Even particle physicists are only dimly conscious of the full meaning of the word 'particle' most of the time. I for one really do think of particles like little marbles when I'm doing my day job. It's a mental picture that works just fine most of the time. But it's wrong.

This simplistic view misses the true complexity, beauty, and downright weirdness of what modern particle physics tells us the world is ultimately made from. This deeper picture is only revealed when we start to think really hard about particles. In the process, we will discover that particles aren't the fundamental building blocks of nature at all. Instead, a new set of objects emerges that are far stranger and less tangible than anything we experience in our everyday lives. These objects are still only partly understood, even by the world's brainiest theorists, but seem to be the true ingredients of our universe.

CREATION AND ANNIHILATION

A few years ago, I worked on a small exhibition at the Science Museum in London, taking up a couple of big showcases in a corner of the bustling Exploring Space gallery. Little noticed among the various bits of physics ephemera by the throngs of screaming children who tear through the gallery like flocks of excitable, high-viz-clad geese was a loosely bound bundle of papers – Paul Dirac's original PhD dissertation. The title, handwritten in charmingly uneven capital letters, is simply 'Quantum Mechanics'. That's one punchy title for a PhD.*

Paul Dirac was one of the most brilliant theoretical physicists of the twentieth century, probably coming second only to Albert Einstein. To give you a sense of his prodigious ability, within just three months of reading the paper where the German theorist Werner Heisenberg first laid the foundations of quantum mechanics, Dirac had produced a brand-new version of the theory, reframing and extending Heisenberg's ideas in a more elegant mathematical language. All at the age of just twenty-three. It's people like that who make you realize how little you've accomplished.

Dirac was also one of physics' strangest figures, and if you've met many physicists, you'll know there's some stiff competition in that department. He was socially awkward, literal minded to a fault, and so uncommunicative that his colleagues defined the unit of a Dirac as one word spoken per hour. Like a visitor from another planet, he struggled to understand common human pastimes, particularly poetry – which he summarized as stating the obvious incomprehensibly – and worst of all, dancing. Among the many Dirac stories recounted by his colleagues was the time he asked Heisenberg why he enjoyed dancing while they were on a scientific jolly together in Japan. After Heisenberg replied that it was a

---

* By comparison, mine was called 'A Measurement of the B_s^0 to K^+K^- lifetime at the LHCb Experiment'. You can guess which one had more impact.

pleasure to dance with nice girls, Dirac sat in thought for several minutes before responding, 'Heisenberg, how do you know *beforehand* that the girls are nice?'

However, despite his inability to get his head around ordinary human behaviour, Dirac was almost without rival when it came to understanding the behaviour of the smallest ingredients of nature. During the first few years of his scientific career he would lay the foundations upon which all of modern particle physics is built. His first step was in making sense of what happens when a photon is born.

Photons are created and destroyed all the time. Every time you flick a light switch or idly tap your phone, you create billions and billions of photons, which then get destroyed almost immediately as they crash into your eyes, the walls of the room, or whatever else gets in their way. Similarly, when an electron orbiting an atom falls from a higher energy level to a lower one, a photon is created that carries off the difference in energy between the two levels. The question is, what is actually happening when a photon is created?

To answer this question, we need to go back to the prequantum view of light, a view that was based on the nineteenth-century concept of the electromagnetic field. We owe the idea of fields in large part to the English scientist Michael Faraday, who spent years getting hands on with electromagnetic phenomena, experimenting with magnets, coils of wire, and dynamos in his basement lab at the Royal Institution in London. In the process, he became convinced that the electric and magnetic forces that he was playing with were communicated by invisible and yet undeniably physical entities – electric and magnetic fields.

Formally speaking, a field is a pretty abstract concept: a mathematical object with a numerical value at every point in space. However, fields are far more than just mathematical abstractions. If you've ever picked up two bar magnets and pushed their north poles toward each other, you'll have felt a powerful force pushing back. Jostle the magnets around a bit and this force changes in strength and direction, as if you were feeling out the edges of some invisible, repulsive thing. You can look as hard as you like at the gap between the two magnets and you won't see anything except empty space, and yet you can feel that there's something there. What you're feeling is a magnetic field, and once you've felt it, it's impossible to deny that it's real.

Faraday found that he could even make magnetic fields visible. Sprinkling iron filings onto a piece of waxed paper placed on top of a magnet, he produced beautiful images tracing out the field's otherwise invisible influence. You can still see Faraday's stunning field maps if you turn up at the Royal Institution on Albemarle Street in London and ask nicely. As the son of a poor blacksmith's apprentice, Faraday hadn't received much in the way of a formal mathematical education, so he instead developed powerful visual representations of his electric and magnetic fields based on lines of force, which he sketched flowing outwards from the north pole of a magnet and back in through the south, or from a positive electric charge to a negative one. You were probably made to draw diagrams like that in school; I certainly was. He imagined these lines as real, physical objects that would move or even vibrate when magnets or electric charges moved, in the same way that you can send a wave rippling down a rope by suddenly flicking one of its ends.

It was the Scottish physicist James Clerk Maxwell who took Faraday's intuitive understanding of electromagnetic phenomena and translated it into mathematical language. In the process he discovered an equation describing a wave of intertwined electric and magnetic fields, dancing through space together. Astonishingly, when he calculated the wave's speed, he found it was exactly the same as the speed of light. Maxwell's theory seemed to show that light was a wave in a unified electromagnetic field.

By the time Dirac was working as a young scientist in the late 1920s, Maxwell's electromagnetic theory of light had been tremendously successful, not least as the basis of wireless communication and radio broadcasting. However, Maxwell and Faraday's electromagnetic field was a continuous object, and it was tricky to see how it could be reconciled with quantum theory, which described light as the flow of individual photons. The challenge was to get the two descriptions of light to play nicely with each other.

Dirac's breakthrough came during a six-month stay at Niels Bohr's Institute for Theoretical Physics in Copenhagen in the autumn of 1926, which followed hot on the heels of Dirac's triumphant PhD dissertation. While Bohr cultivated an open, relaxed atmosphere where lively discussion was encouraged, Dirac preferred to work alone, locking himself away in the library during the daytime and taking long solo walks around the city after dark. When he did attend the discussion sessions, he would sit listening in silence, responding when prompted with a monosyllabic yes or no. His colleagues, Bohr included, didn't know what to make of the strange Englishman.

It was probably during one of his solitary days in the institute's library that Dirac started to ponder the thorny issue of making photons. As a physicist working at the beating heart of the quantum revolution, you might have expected Dirac to take light quanta as his starting point and try to build the electromagnetic field from a multitude of tiny photons, in much the same way an ocean is made up of vast numbers of individual water molecules. But that isn't what he did. Instead, Dirac took the electromagnetic field as the fundamental thing. It was photons that were made of the electromagnetic field, not the other way around. A photon, said Dirac, was nothing more than a discrete, transitory little ripple in the ever-present electromagnetic field.

Dirac had just invented a brand-new physical entity, a 'quantum field' – a strange amalgam of Faraday's electromagnetic field and Einstein's photons. In many ways the quantum electromagnetic field looks a lot like Faraday's ordinary nonquantum or classical version. Both are invisible and yet fill all of space, can transmit electric and magnetic forces, and, if you wobble them around in the right way, can sustain waves that travel through the field in the form of light. However, there is a crucial difference between Dirac's quantum field and the old classical one. Whereas you can create a wave of any size you like in the classical electromagnetic field, in quantum field theory, there is a fundamental minimum amount of waviness that you can have. This is what we call a 'photon'.

To try to understand this a bit better, imagine two friends, let's call them Alice and Bob,* standing a few metres apart, each holding one end of a tightly stretched length of elasticated bungee cord. In this analogy, the one-dimensional bungee cord is standing in for the admittedly three-dimensional electromagnetic field, but let's not overcomplicate things. Now imagine that Alice starts to move her end of the cord up and down at a rate of, say, three wobbles per second, while Bob keeps his end still. As she moves her hand, waves start to ripple along the length of rope until they reach Bob at the other end. Now, as this is an ordinary, classical bungee cord, Alice can choose to move her hand up and down by any amount she likes; she can make little waves that are 5 centimetres high or wave her hand around wildly and make waves that are as high as she is, or any size in between. This is pretty similar to how light waves are created in a classical electromagnetic field; you just need to substitute Alice's hand for a charged particle like an electron.

Now let's say we give Alice and Bob a quantum bungee cord (to be clear, there is no such thing, but go with it). Because the cord now obeys the laws of quantum field theory, Alice discovers something strange. She can no longer create waves of any height she likes. If she moves her hand up and down, still at a frequency of three wobbles per second, by say 5 centimetres, the bungee cord remains mysteriously still. Try as she might, she cannot make a wave 5 centimetres high, or indeed 6 centimetres, 7 centimetres, or

* Stars of many a physics analogy, who first appeared as fictional characters in Ron Rivest, Adi Shamir, and Leonard Adleman's 1978 paper on cryptography.

8 centimetres. However, when she moves her hand up and down by 10 centimetres suddenly a wave pings along the cord, perhaps startling Bob out of his daydream at the other end. There seems to be a fundamental minimum amplitude that a wave can have on this quantum bungee cord. In the electromagnetic field we'd call this smallest possible wave a photon, so I guess in this analogy a quantum of the bungee cord might be called a 'bungeeon'.

The same is true for the quantum electromagnetic field. For a given frequency of light, you can only add energy to the electromagnetic field in discrete little lumps. The field can have no photons, one photon, two photons, or a quadrillion photons rippling about in it, but you can't have a bit of a photon. They must come in whole numbers – or, to put it more scientifically, the electromagnetic field is 'quantized'.

Dirac described the process of creating and annihilating photons in rather more abstract terms, inventing mathematical objects called 'creation operators' and 'annihilation operators'. As the names suggest, the creation operator injects one photon into the electromagnetic field, while the annihilation operator takes one out. Using this mathematical language, Dirac was able to calculate how likely an atom was to absorb or emit a photon in certain circumstances, finding an answer that agreed perfectly with a more ad hoc calculation performed by Einstein ten years earlier.

Dirac's quantum field theory was a triumph; not only had he gone one better than Einstein, he also believed he had laid to rest all the handwringing over wave–particle duality.* There was no longer any need to think of photons sometimes as waves and sometimes as particles; instead they could be understood as vibrations of a single unified object, the quantum electromagnetic field.

However, Dirac's theory was only half the story. Perhaps you can think of photons as little ripples in an electromagnetic field,

---

* He hadn't. Today there's an entire community of researchers who spend their time thinking about this stuff.

but what about the particles of matter? Electrons and protons seemed to be rather different beasts. Sure, they exhibit the same wave–particle duality as photons, but as far as anyone could tell, it was impossible to create or destroy them. Unlike photons, which blink in and out of existence willy-nilly, electrons and protons seemed to be eternal.

To understand the birth and deaths of matter particles we need to bring in the other great revolutionary theory of the early twentieth century, special relativity. Just as quantum mechanics had upended the laws governing atoms and particles, special relativity redefined what we mean by space and time, with some delightfully counter-intuitive results. At its core is the principle proposed by Einstein that the laws of physics – and, crucially, the speed of light – are always the same regardless of how fast you are moving. It turns out that to make this work, you have to be willing to let go of a universal definition of space and time that everyone can agree on. Instead (for reasons too thorny to get into here) space and time become *relative*, with the distances that we measure between objects or the number of ticks of a clock between two events depending on how fast we are moving relative to one another.

The versions of quantum mechanics doing the rounds in the mid-1920s were inconsistent with special relativity. In other words, two observers moving at different speeds would disagree on what the laws of quantum mechanics were. It was clear that this meant quantum mechanics was, at best, incomplete, but fusing it with special relativity proved to be no easy task.

Over the summer of 1926, about six different physicists thought they had found an equation that might do the trick. It is known as the Klein–Gordon equation (after two of its discoverers, Oskar Klein and Walter Gordon) and it appears to describe the quantum behaviour of an electron travelling close to the speed of light in a way that is consistent with the edicts of special relativity. In particular, it includes special relativity's most famous consequence – the equivalence of mass and energy as captured by $E = mc^2$ – by including the mass-energy of the electron as a term in the equation.

Niels Bohr, for one, thought the problem of getting a relativistic equation for the electron was solved. Dirac, though, was unconvinced. For one thing, the wave functions of the Klein–Gordon equation couldn't easily be interpreted as the probability of finding a particle in a particular location as it could in ordinary quantum mechanics. Dirac was sure he could do better.

Back in Cambridge in the autumn of 1927 after a few months spent in the charming medieval German university town of Göttingen with the quantum power trio of Max Born, Werner Heisenberg, and Pascual Jordan, Dirac attacked the problem with quiet determination. No longer a lowly PhD student but a fellow of the grand St John's College, Dirac now had his own comfortable, albeit rather spartan, rooms in the college's picturesque grounds on the banks of the River Cam. As usual, he worked alone, scribbling out pages of algebra at his little desk from early in the morning until dusk, only taking breaks on Sundays when he would go for long walks in the Cambridgeshire countryside, climbing the occasional tree, all dressed in a full three-piece suit.

Dirac knew that he was unlikely to be able to derive the relativistic equation for the electron from some profound universal principle in the way Einstein had derived relativity. Instead, as is often the case in physics, he would have to make a series of educated guesses. There were, however, some features he knew the equation needed to have that he could use as guides. First off, it had to be consistent with special relativity, which meant it needed to look the same regardless of how fast an observer was moving and also to include the mass-energy of the electron. Secondly, at speeds much lower than the speed of light, the equation should look like ordinary bog-standard quantum mechanics. Finally, to make sure the wave function of the electron could be straightforwardly interpreted in terms of probabilities, he was convinced that the equation needed to be 'first order' in space and time – in other words, contain space and time just as they are, rather than squared (second order) as they appeared in the Klein–Gordon equation.

After months of guessing, testing, and discarding possible equations, Dirac finally came upon a promising-looking candidate. Not only did it agree with both relativity and quantum mechanics, it also naturally accounted for a hitherto mysterious property of the electron, the fact that it behaves as if it is spinning.* Solving his equation, Dirac found two solutions, one describing the electron with its spin pointing up, the other with its spin pointing down. The spin of the electron emerged almost miraculously from Dirac's unification of quantum mechanics and special relativity. If the electron's spin hadn't already been discovered in experiments, his equation would have predicted it.

Dirac was elated. Not only had he pulled off one of the greatest feats in the history of theoretical physics, he had discovered an equation of near-incomparable beauty. The concept of mathematical beauty is a bit hard to define, although many mathematicians will recognize it when they see it, a bit like you might recognize beauty in the smooth, clean lines of a sailing ship. Dirac's equation had a piercing simplicity, simultaneously resolving several intractable problems while employing an absolute minimum of extraneous bells and whistles, like a razor-sharp blade cutting through a dense tangle of undergrowth. The equation has what some theorists would describe as a feeling of inevitability, the sense that it is so simple, so elegant, and yet so powerful that it could not have possibly been otherwise. I am now going to commit a cardinal sin of popular science writing and show you the equation I am babbling on about:

$$(i\gamma^\mu\partial_\mu - m)\psi = 0$$

---

* All matter particles, including the electron, have total spin ½, which can be pointing either 'up' (spin +½) or 'down' (spin −½).

Isn't that a gorgeous thing? Even if the sight of algebra makes you feel dizzy, I hope I can impress upon you just how lean and mean this equation is.* It has only three bits – the first term, $i\gamma^\mu\partial_\mu$, describes how the electron changes through space and time, the m is its mass, and finally ψ is the electron's wave function (the mathematical object that tells you the probability of finding the electron in a particular place or state) – and yet despite this simplicity it describes every electron that ever was or ever will be.

Dirac kept his monumental discovery quiet for more than a month, suffering occasional bouts of intense panic at the thought that his beautiful equation might unravel when it was forced to confront experimental reality. Fearing the result, he kept putting off checking whether it could accurately predict the energy levels of the hydrogen atom, a test he knew the equation had to pass. However, when he did eventually bring himself to do the maths, he found that it not only got the answer right, it actually matched the experimental data even more closely than ordinary quantum mechanics.

When Dirac finally allowed his equation out into the wild at the start of 1928, it set the physics world on fire. His rivals in continental Europe's theoretical powerhouses reacted with a mixture of wonder and dismay. Pascual Jordan, who had been working on the same problem, was left totally demoralized, while Heisenberg spoke of an English physicist who was so clever there was no point trying to compete with him.

However, at the back of Dirac's mind was a gnawing anxiety; he suspected that there was something deeply wrong with his equation. He had discovered that it had not two but four different solutions. The first two were all fine and dandy, describing the established spin up and spin down states of the electron, but the

---

* To be fair, this version of the equation uses a more compact notation than the slightly more intimidating version that Dirac first wrote down, but the physics and the structure of the equation are identical.

other two seemed to describe something profoundly disturbing – electrons with *negative energy* (not to be confused with negative charge).

The idea of negative energy electrons makes about as much sense as a pond with a negative number of ducks. At first, Dirac was tempted to sweep those solutions under the carpet, but he soon realized that they couldn't be ignored so easily. If these negative energy states existed, then ordinary positive energy electrons should be able to fall into them, like a pool ball tumbling into a pocket.

The problem was that no one had ever seen an electron falling into a negative energy state. Determined to save his beautiful equation, Dirac proposed a rather brazen solution: the reason we never see electrons falling into negative energy states is that the negative energy states are already full. An electron trying to jump from a positive to a negative energy state finds its way blocked by an existing electron, like a ball kept out of a pocket by a stack of previously potted balls.

In principle, this means that the entire universe is filled by an infinite sea of negative energy electrons. This raises an obvious question: why don't we notice them? Surely it'd be pretty obvious if we spent our lives wading through an infinite number of electrons? Not necessarily, said Dirac. As long as these negative energy electrons were perfectly evenly distributed throughout space, then they would fade into the background.

This electron-sea solution didn't end Dirac's woes. What if, for example, a photon were to crash into one of these negative energy electrons and kick it up into a positive energy state? Suddenly, we'd see an electron appear out of nowhere as it emerged from below the waterline. Meanwhile, an electron-shaped hole would be left in the sea, spoiling the perfect uniformity that kept it hidden. However, rather than think of this as a hole in an infinite sea of negatively charged, negative energy electrons, Dirac realized that the hole would behave like a positive energy electron – but one with a *positive* charge.

Therein lay the rub; according to all experiments to date, there was no such thing as a positively charged electron. Every electron that had ever been seen had been negatively charged. At first, Dirac tried to show that these positively charged holes might in fact be protons, but you'd expect the hole to have the same mass as the electron, and a proton weighs almost two thousand times as much. Worse still, if protons really were holes in the negative energy sea, then electrons should be able to fall into them, annihilating both electron and proton and leading to the instantaneous destruction of every atom in the universe.

Despite these problems and the gloom of many of his colleagues, Dirac's conviction of the beauty and rightness of his equation was unshakeable. By 1931, after all attempts to get rid of the negative energy states had failed, he was ready to make his most audacious prediction: positively charged electrons really must exist in nature.

What happened next still gives me goosebumps. A year later, thousands of kilometres away in California, a positively charged electron turned up in a cloud chamber photograph taken by the young American experimental physicist Carl Anderson, who had been studying cosmic rays raining down from the heavens. Hot on the heels of Anderson's paper, the Cavendish physicists Patrick Blackett and Giuseppe Occhialini spotted more positive electrons, this time popping into existence accompanied by ordinary negatively charged electrons when a cosmic ray smashed into an atom in their cloud chamber. Dirac, a frequent visitor to the Cavendish Laboratory, which in those days was still ruled by the booming Ernest Rutherford, was soon poring over the photographs with Blackett, making calculations and checking the results against his equation. It didn't take long to realize that these positive electrons were the very same particles that Dirac had predicted must exist.

Dirac had achieved something truly miraculous. Using the power of pure thought, he had conjured the existence of a form of matter that had never been seen before in nature. By bringing together quantum mechanics and special relativity and following his nose, he had opened a window into the world of antimatter, a mirror image of the ordinary stuff that makes up the visible universe. We now know that every matter particle has an antiversion, with precisely the same properties but opposite charge. Dirac's positive electron is now known as the positron or the antielectron. Meanwhile, the proton has a negatively charged version, the antiproton, and there are also antineutrons, antimuons, antiquarks, and antineutrinos. The fact that Dirac managed to predict such a fantastical thing just by thinking really, really hard has surely got to rank as one of the most incredible feats in the history of science.

What's more, the discovery of antimatter destroyed the idea that matter is eternal. Matter particles could now be created by smacking one particle into another with enough energy to make a new particle–antiparticle pair. And the reverse is possible too – if a particle is unlucky enough to meet up with its antiparticle, they annihilate each other, disappearing into oblivion with a flash of radiation.

Of course, this does raise a question: if matter and antimatter are always created and destroyed together, how come the universe is made only of ordinary matter? As we'll see, this rather troublesome conundrum will come back to bite us.

There is one bit of Dirac's work that hasn't stood the test of time: the idea that antiparticles are holes in a negative energy sea. Within a few years, physicists found a way to do away with Dirac's sea altogether by describing electrons and positrons in the same way as photons – as vibrations in quantum fields. The boundaries between fields and particles, light and matter, had finally dissolved.

Today, we physicists think of all particles this way. For every particle we've met along our journey so far there is a corresponding quantum field. Photons are little ripples in the electromagnetic field; electrons and positrons, likewise, are ripples in something called the 'electron field'. Up quarks are little ripples in the up quark field, and so on and so on. When two protons smash into each other at the LHC, they set the quantum fields of nature ringing like bells, sending a cascade of ripples outwards through our detectors, each a different musical note in a quantum mechanical symphony. We interpret these ripples as particles, but what we believe we are really seeing are transitory wobbles in quantum fields.

In fact, you might even say that there's no such thing as a particle. As far as we can tell, the real building blocks of the universe are quantum fields: invisible, fluid-like substances that we can't see or taste or touch, and yet are all around us, stretching from deep within the smallest atom of your being to the farthest reaches of the cosmos. Quantum fields – not chemical elements, nor atoms, nor electrons, nor quarks – are the real ingredients of matter. We are walking, talking, thinking bundles of tiny self-perpetuating disturbances sloshing about in intangible quantum fields.

Of course, things aren't quite that simple. While it would be lovely and comforting if we could just think of an electron as a little ripple in the electron field, that is really only half the story. It turns out that even an object as simple as an electron is a fantastically complex thing, not merely a ripple in the electron field but a baroque mixture of *every* quantum field in nature. While this makes calculations in quantum field theory fiendishly difficult, it also opens up opportunities to explore nature in ways that would be totally impossible in either quantum mechanics or special relativity alone. In particular, experiments that study the electron in exquisite detail have the potential to teach us about both the electron itself and even quantum fields that we have never seen before. One such experiment is going on right now under the bustling streets of London.

DRESSING THE ELECTRON

Squeezed into a pokey basement lab at Imperial College in central London is an experiment that can pull off the same trick as the Large Hadron Collider for a thousandth of the price. Just metres beneath the thunder of London traffic and the ceaseless hammering of footsteps from the thousands of tourists and schoolchildren who throng to the museums of South Kensington, a small team of physicists are making one of the most delicate measurements of a fundamental particle ever attempted.

Their mission is to measure the shape of the electron. The idea that a fundamental particle can have a shape might seem strange, particularly given that we just said that particles are shape-shifting ripples in quantum fields, but park that thought for just a moment. The really surprising thing is that by measuring the electron's shape with fantastic precision, the team at Imperial can search for hints of quantum fields that we've never seen before, potentially uncovering evidence for particles with masses so huge that even the LHC wouldn't be powerful enough to produce them.

How on earth can measuring the shape of a puny little electron tell us anything about particles with gigantic masses? Well, it all comes down to the fact that particles are really just ripples in quantum fields, a fact that has dramatic consequences for the properties of an electron. To understand properly what the physicists at Imperial are up to we need to have a serious think about what an electron really is, and perversely, perhaps, the best way to get started is to consider what quantum field theory tells us about empty space, or what physicists refer to as 'the vacuum'.

Imagine taking a little region of space and sucking out all the atoms, all the particles, every last stray photon and neutrino. What's left? If there are no particles then presumably the answer is nothing at all. Actually, quantum field theory tells us that this little region of 'empty' space is still an amazingly crowded place; it's chock-full of quantum fields. There might not be any particles left, but the fields that they are ripples in are always there. In the standard model of particle physics there are dozens of fields (the exact number depends on how you choose to count them, but for the sake of argument let's say there are twenty-five), including the electron field, the neutrino fields, the quark fields, the electromagnetic field, the gluon fields, and more besides. All these fields exist everywhere, even in the vacuum. Empty space is far from empty.

Now let's say we dump enough energy into the electron field to create a little quantized ripple – a single electron. Since the electron has an electric charge, it has a direct effect on all the quantum fields hanging around in the vacuum. The most obvious thing that happens is that the charge of the electron warps the shape of the electromagnetic field in the area around the electron. Close to the electron the electromagnetic field becomes strong, while farther away from the electron it becomes weaker, eventually fading away to (almost) zero. In principle, this distortion in the electromagnetic field contains some energy, so when we think of an electron we should really consider both the ripple in the electron field *plus* the distortion it creates in the electromagnetic field.

But it doesn't stop there. Since the electromagnetic field is what communicates the electromagnetic force, it is 'connected' to every other quantum field that has an electric charge. This means that the distortion the electron creates in the electromagnetic field causes yet more distortions in a whole bunch of other fields, including the electrically charged quark fields. Now quarks have this property we call 'colour', which means they interact with the gluon fields (the fields of the strong force), and so the distortions in the quark fields cause further distortions in the gluon fields. There is even a back reaction where the distortion in the electromagnetic field causes a further distortion in the original electron field. And so it goes on and on. The upshot of all this is that an electron is not simply a ripple in the electron field; it is a ripple in the electron field *plus* distortions in every quantum field we have ever discovered. What we might call the bare electron – the pure ripple in the electron field – is dressed up in an elaborate gown woven from every quantum field in nature.

The way that quantum fields dress up the electron (and every other particle for that matter) makes calculating even simple processes in quantum field theory very complicated, but on the other hand it gives us a fantastic opportunity to search for the influence of quantum fields we've never seen before. As we'll see in the coming chapters, there are lots of good reasons to believe that there are more quantum fields than the twenty-five-odd we've discovered so far. A good example is dark matter, a mysterious substance that astronomers and cosmologists have shown must be about five times more common than the ordinary atomic stuff that you and I, and every star and planet in the sky, are made from. It's usually assumed that dark matter is some kind of particle, in which case the vacuum should also contain an extra quantum field, the quantum field that dark matter particles are ripples in.

The brute force approach to search for dark matter particles at the LHC is to smack two protons into each other really, really hard and hope that the collision has enough energy to set off a vibration in the dark matter field. If we're lucky, and the amount of energy needed to set the dark matter field wobbling – in other words, the mass of the dark matter particle – is within reach of the LHC, then we should be able to detect evidence of dark matter particles zipping out from the collisions. However, if the mass of the dark matter particle is higher than the maximum energy of the LHC, we won't be able to get a vibration going in the dark matter field, and dark matter will remain a mystery.

There is, however, another way. It relies on how quantum fields dress up fundamental particles. If a quantum field for dark matter exists and it interacts with at least one of the other quantum fields in the standard model, then in principle it should also contribute to the elaborate quantum field gown worn by the electron. If you think of this gown as a fabric woven from threads of different quantum fields, then a few of the threads in the electron's gown would be made of the dark matter quantum field. Since what we measure in experiments is the bare electron, plus its outfit, then if we make superbly precise measurements of the electron, we might be able to detect the subtle effects of *new* quantum fields woven into its quantum mechanical glad rags.

This is exactly what the team at Imperial are trying to do. I first read about their work back in 2011 shortly after the LHC had fired up for its inaugural run. At first glance, what they were claiming seemed impossible; in a small lab in central London with a budget of millions, not billions, they were ruling out the existence of the same quantum fields that thousands of physicists at the world's largest experiment were in the middle of searching for. Since then I had always wanted an excuse to pay a visit and get a glimpse of the miraculous machine that surely lurked beneath the streets of South Ken.

On a crisp February morning, I arrived at Imperial College's Blackett Laboratory, a 1960s block plonked unselfconsciously amid the neighbourhood's Victorian grandeur, just across the road from the Royal Albert Hall. I was met in the foyer by Isabel Rabey and Sid Wright, two young postdoc researchers who have dedicated years of their lives to the experiment. Isabel was back in town to see her old team, having spent her PhD down in the basement improving the experiment before heading off for a position at the Max Planck Institute near Munich, while Sid, a relative newcomer, had taken on a lot of the work at the coal face following Isabel's departure. They seemed pleased to have an opportunity to talk about what is obviously a labour of love. I explained how as an LHC physicist I had been keen to see the incredible experiment that was giving us a run for our money. Isabel laughed. 'I think you're in for a shock.'

Down a couple of flights of an echoing stairwell and along a short corridor, they welcomed me into their laboratory. It really was tiny, hardly any bigger than the living room of my not-very-big London flat. Isabel told me that when they showed around representatives of the national agency that funds large particle physics and astronomy projects in the United Kingdom the visitors had been taken aback by just how pint-sized their experiment was. 'We got the sense that they were thinking, "If you made it a bit bigger, perhaps we could fund you."'

To see the experiment properly we had to shuffle one by one between a wall and a heavy piece of shielding protecting the experiment from any stray magnetic fields that might interfere with the delicate measurements. To our right was a bank of oscilloscopes and electrical gizmos that were used to read out and monitor the experiment, on the left a large table covered with optical elements that glowed with lurid green laser light, and in the centre was the business end of the experiment, which, I hope Isabel and Sid won't mind me saying, looked to my untrained eyes like a metal dustbin.

It was certainly a far cry from the towering experiments of the Large Hadron Collider, which one journalist I showed around compared to the giant alien portals from the 1990s sci-fi show *Stargate*. If the LHC is Stargate then Isabel and Sid's experiment was more like Doc Brown's time-travelling DeLorean from *Back to the Future*, a little ramshackle-looking perhaps, but remarkably effective.

Isabel and Sid gamely talked me through how the experiment worked. I will readily admit that my atomic and molecular physics is a bit rusty, and I felt like a slowcoach as I tried to get my head around the complex system of lasers and magnetic and electric fields used to measure the shape of the electron.

The first thing to understand is what we mean by the shape of the electron. Strictly speaking, the experiment measures something called an 'electric dipole moment' (EDM), which is a measure of how the electric charge of the electron is spread out in space. An EDM of zero means that the electron's charge is distributed in a perfectly symmetrical sphere, while a nonzero EDM would mean that the electron is like a cigar, more negatively charged at one end and more positively charged at the other. The electron's EDM turns out to be extremely sensitive to whatever quantum fields are hanging around in the vacuum in the vicinity of the electron, which is why it's such an interesting quantity to measure. If you do some serious number crunching using only the quantum fields that we already know about you find that the EDM of the electron should be stupendously tiny, with a value of 10^{-38} e cm (an e cm is the unit of the electric dipole moment, where e is the charge of the electron and cm is a centimetre, but don't worry about the details; the main thing is that 10^{-38} is really, really small). This is so tiny that if there are no new quantum fields beyond the ones that we already know about, then the electron should appear perfectly spherical to any experiment that we can currently imagine.

However, many popular theories that attempt to explain dark matter and other mysteries introduce new quantum fields that dress the electron in a way that squashes it into a far more pronounced cigar shape, in some cases increasing its EDM by a factor of more than a trillion. Such huge enhancements would put it within reach of the Imperial experiment, allowing the team to potentially discover hints of new quantum fields – without the aid of a 27-kilometre particle collider.

But even with a huge boost from new quantum fields, the electron's EDM would still be indescribably minute, and measuring it requires a correspondingly ingenious experiment. Rather than measure electrons directly, the team at Imperial study ytterbium fluoride, a molecule of the rare metal ytterbium and fluorine gas, carefully chosen thanks to its special sensitivity to the electron's EDM. In particular, the outermost electron in an ytterbium fluoride molecule can exist in two different energy levels, one where the electron's spin is pointing up and the other where it's pointing down. The crucial point is that energies of these up and down levels are shifted in *opposite* directions by the EDM of the electron. In other words, if the electron has a large EDM, then one level gets shifted up in energy, while the other gets shifted down. So if you can measure the difference in energy between these two levels then you can indirectly measure the electron's EDM. This property means ytterbium fluoride acts a bit like a magnifying glass that makes you a million times more sensitive to the electron's EDM than if you tried to measure it using electrons flying around outside the confines of an atom or molecule.

Working with these molecules comes at a cost, including the fact that ytterbium fluoride is so unstable that you have to create it continuously inside the experiment. As we stood in the lab, Sid pointed out a continuous *drr-drr-drr-drr-drr-drr*, the sound of a laser striking an ytterbium metal target twenty-five times per second, vaporizing little puffs of ytterbium from a solid metal block, which then react with fluorine gas to form tiny clouds of ytterbium fluoride molecules. A clever system of lasers, radio waves, and microwaves then puts the molecules into a mixture of the spin up and spin down energy levels before they are allowed to drift upwards through the metal cylinder (the thing I compared to a dustbin), which contains an electric field.

The up and down states get opposite energy shifts from the electric field, and the size of their shifts depends on the size of the electron's EDM – the bigger the EDM the bigger the energy shift. Once the molecules exit the electric field at the top of the cylinder, they are measured using a laser, allowing Isabel, Sid, and the team to determine the energy shift and, after painstaking months of data collecting, measure the electron's EDM.

That's the idea at least. In practice making the measurement is extremely tricky. The instrument is so sensitive that it can be affected by all kinds of external influences. Particularly troublesome are stray magnetic fields. They once discovered that a problem they were having with the experiment was due to a powerful magnet being used by a different team two floors above them. A stern word from their boss and originator of the EDM experiment, Professor Ed Hinds, soon saw the other team agree to move their magnet to a higher floor. Sid also told me that they'd noticed that the magnetic interference gets much worse when London Underground trains are running.*

---

* Apparently, the Piccadilly Line is the worst culprit.

The Imperial team led by Ed Hinds released their first measurement back in 2011, finding that the electron appeared to be exquisitely round, all the way down to a precision of 10^{-27} e cm. To give you a sense of just how spherical that is, if you were to blow up an electron to the size of the solar system, it would be spherical to within the width of a single strand of human hair!

Disappointingly (at least for particle physicists), the marvellous roundness of the electron ruled out the existence of a bunch of new quantum fields that I and my colleagues at the LHC were busily searching for at the time. However, the measurement itself was a real tour de force; not only was it the most precise in the world, it was also the first time that molecules, rather than atoms, had been used in an EDM measurement. Back then, every other experiment used single atoms, and many of the Imperial group's rivals had thought they were wasting their time trying to make the delicate measurement work with relatively messier molecules. Nowadays, though, almost all their rivals are following the trail blazed by the Imperial team, thanks to molecules' powerful EDM magnifying properties.

Indeed the ACME experiment, run by a joint Harvard–Yale team in the United States, has since leapfrogged the Imperial measurement, pushing the EDM down by another factor of a hundred, with a second team based in Colorado snapping at their heels. To catch up, the Imperial team are now developing a new secret weapon, an upgraded version of their experiment, which I was given a glimpse of in a neighbouring, far more spacious lab. 'This one'll look more like what you're used to at CERN,' Sid assured me as we entered. In front of us was a gleaming stainless-steel tube, not unlike a particle accelerator, which will eventually be extended to span the full length of the laboratory. A longer tube means that the molecules spend more time in the electric field, giving a bigger energy shift and a big boost to the experiment's sensitivity. When their new instrument starts collecting data, it'll effectively be probing quantum fields whose particles could have masses well above what the LHC is able to produce directly, providing a fantastic opportunity to discover more of nature's fundamental ingredients.

We've certainly come a long way from that first apple pie experiment. Back then we were dealing with tangible things you could taste and touch: jagged lumps of black carbon, oily liquids, curling tendrils of acrid vapour. The ingredients that we're now left with are about as far from tangible as you can get – invisible, ethereal, omnipresent quantum fields. The apparent solidity of the world turns out to be an illusion, a conjuror's trick. There are no indivisible atoms as the ancients thought. Democritus, for all his ancient beardy wisdom, was wrong. Nature, deep down at its roots, is continuous, not discrete. I've used the phrase 'building blocks' throughout this book to talk about the fundamental ingredients of nature, but in truth there are no such things. The apparent 'blockiness' of matter dissolves when we look closely enough. Particles are not particles, they are passing disturbances in quantum fields, entities that strain the imagination and yet fill every last cubic centimetre of the cosmos. All objects – apple pies, humans, stars – are agglomerations of vast multitudes of these vibrations, moving together in a way that creates the illusion of solidity, of permanence. What's more, since there is only one electron field, only one up quark field, and only one down quark field, you and I, dear reader, are connected to each other. Each of our atoms is a ripple in the same cosmic ocean. We are one with each other, and with all of creation.*

At the start of the chapter, we said that the ingredients of an apple pie are electrons, up quarks, and down quarks. Quantum field theory tells us that these three particles are vibrations in three

* At the risk of getting a bit too Neil deGrasse Tyson, that also means that we are all one with lots of unpleasant stuff: the Ebola virus, dog shit, and Piers Morgan, for instance.

corresponding fields. However, as we just saw, even this is a gross oversimplification. An electron is not merely a ripple in the electron field, but a complicated mixture of distortions in every quantum field that we have ever discovered. The same goes for the up quarks and down quarks that make up protons and neutrons in the nucleus. This means that to fully understand the make-up of our apple pie we need to know about every last quantum field in nature, even those whose particles are too unstable or weakly interacting to bind together to make atoms.

Our current best description of the known quantum fields is the standard model of particle physics, an exceptionally bland name for one of the greatest achievements of human thought. We have already met many of its stars: the electron, the quarks, neutrinos, gluons. However, there is one key piece of this picture that was found only in the last few years, the final ingredient of our apple pie, and one that opens up a Pandora's box of new problems and opportunities.

CHAPTER 10

The Final Ingredient

I first arrived at CERN, the European Organization for Nuclear Research on the outskirts of Geneva, on a sunny afternoon in July 2007. I was a fresh-faced, twenty-one-year-old undergraduate with an unspoiled enthusiasm for physics and what in hindsight was rather ill-advised shoulder-length hair. For a few weeks that summer I and more than a hundred other summer students from across Europe would get a taste of life at the cutting edge of particle physics.

The CERN of my imagination was a gleaming, futuristic place lifted straight from the pages of science fiction, where Promethean scientists used gigantic subterranean machines to probe the very nature of reality. I was therefore a little taken aback to find myself deposited at the gates of what looked more like a scruffy 1960s university campus in need of some serious TLC: a chaotic jumble of shabby office buildings and dilapidated warehouses with peeling paintwork and rusty corrugated roofs.

Many first-time arrivals at CERN experience a similar culture shock, a milder cousin of so-called Paris syndrome, which affects some visitors to the City of Lights when they find themselves in a far grubbier, noisier, and ruder place than the fairy tale of books and movies. On the other hand, while Paris syndrome can apparently induce symptoms as extreme as paranoia, dizziness, and hallucinations, CERN syndrome only left me with a vague feeling of disappointment.

A lack of sci-fi whizziness notwithstanding, I had arrived at the lab at one of the most exciting possible moments. After three decades of planning, fundraising, and construction, the Large Hadron Collider, the world's biggest machine, was just months from firing up for the very first time. I was to spend the summer working on the CMS experiment, one of the four cathedral-sized detectors whose job would be to scour the collisions produced by the LHC in search of new fundamental particles.

My specific project was to help get one of CMS's subsystems ready to collect data. In practice this involved sitting in an office staring in puzzlement at reams of computer code, a task I was woefully unprepared for – no one had thought to teach us any coding at university.* To make matters worse, the person who was supposed to be supervising me was away for my first two weeks, which only compounded my sense of being lost in a strange, dreary world.

Then, two weeks in, everything changed. One afternoon, some fellow students and I were taken by minibus to the far side of the 27-kilometre LHC ring to visit the site of the CMS experiment. We pulled up at a fenced compound surrounded by peaceful French farmland, at the centre of which was a large hangar-like building. What I saw inside took my breath away. Towering above us were huge sections of the detector, each more than three storeys high, lined up ready to be lowered down an enormous concrete shaft at the far end of the hangar, which plunged vertically a hundred metres through the earth to the experimental cavern below.

At last, here was the sci-fi magic I had been waiting for. Most mesmerizing of all was the huge object farthest from the access

---

* To be fair, we did learn how to measure the strength of gravity by rolling a ball down a slope, which actually turns out to be excellent training for a career in physics (in the seventeenth century).

shaft, the piece of CMS that would be lowered down last and slid into place to complete the experiment – the so-called endcap. Unfortunately, there are no handy comparisons to help you picture what the endcap looked like – it was such an otherworldly thing – but imagine, if you can, a twelve-sided disc balancing on its narrow edge, 15 metres top to bottom and side to side, roughly the size of three double-decker buses stacked on top of one another. Its red surface was criss-crossed by bright blue cabling and at the centre of the disc was a large black and silver cylinder protruding outwards like the hubcap of some giant alien wheel.

Seeing these monumental slabs of the detector laid out waiting to go underground suddenly made the whole enterprise much more real to me. Standing beneath them, you began to properly appreciate the decades of work that were involved in making this experiment a reality. Every tiny component had been painstakingly researched, designed, built, and tested before being shipped to CERN from labs all over the world, and tested again before finally being installed.

But the best was yet to come. After several minutes wandering around the hangar, gawping at the huge slices of detector, we took a lift down to the experimental cavern itself. Standing on a metal gantry raised 10 metres above the cavern floor you could survey the near-completed experiment. CMS stands for Compact Muon Solenoid, which I've always thought is rather a strange use of the word 'compact'. The detector is shaped like a giant barrel lying on its side, 15 metres high and 22 metres long and weighing in at a whopping 12,500 tonnes. The whole thing contains enough iron to build two Eiffel Towers.*

The protons collide at the centre of this huge barrel, sending particles flying out through concentric subdetectors, like the layers of an onion, each of which provides different information about

---

* Compact, I guess, is a relative term. Its rival experiment, ATLAS, on the opposite side of the ring, is almost twice the height, width, and length of CMS.

the particles: their momentum, energies, directions, and so on. My specific project involved helping to get one of these layers, known as the 'electromagnetic calorimeter' (ECAL), ready for data collecting.

The ECAL is one of the star features of CMS, a gleaming crystal cylinder made up of more than 75,000 transparent blocks of lead tungstate, a compound of lead, tungsten, and oxygen that looks like glass but has the heft of a block of lead. When an electron or photon smacks into one of these crystal blocks, it releases a burst of light, which is recorded by a sensor glued to one end, allowing the energies of the photon or electron to be measured.

The challenges of building the ECAL were immense. Each crystal had to be grown slowly over a period of two days in specially designed platinum-lined crucibles, and the number needed was so huge that a former Soviet military factory had to be recommissioned for the purpose, which along with a second facility in China spent ten years churning out crystals, day in, day out. What's more, the Chinese facility didn't have access to the quantities of platinum required to line the crucibles, and so the CMS management team had to go to the vaults of UBS in Zurich and borrow $10 million worth of platinum on the promise that it would be returned once the crystals were finished.

You might wonder why you would go to such great lengths to build just one component of the detector. Well, the ECAL has an absolutely critical role in the experiment: its job is to measure the energies of photons and electrons, and it needs to be able to do this fantastically precisely. Why? Well, it all comes down to the raison d'être of the Large Hadron Collider: to find the last missing piece of the standard model of particle physics, a particle whose existence would finally explain the origin of two of the forces of nature and why fundamental particles have mass. In fact, it was so important to our understanding of the laws of nature that it became, rather hyperbolically (and unhelpfully), known as the 'God particle'. I am speaking, of course, of the Higgs boson.

Back in the 1990s, the physicists planning CMS realized that one of the best ways to spot signs of the Higgs would be via its decay into two high-energy photons. As is the case with most new particles discovered at colliders, you could never hope to detect a Higgs boson directly. If one got produced by a collision at the LHC, it would decay into other particles almost as soon as it came into being, living far too short a time to reach the sensitive parts of the detector. Instead, all you would detect would be the particles it decays into. Physicists would then have to wade through huge quantities of data in search of particles that might have been produced by the decay of a Higgs, and its decay into two photons would be the easiest to spot. Easy being a relative term, of course.

Difficult to build as CMS was, the physicists knew their crystal electromagnetic calorimeter would give them the best possible chance of snaring the Higgs's telltale decay into two photons. As I stood in that hangar in July 2007, the gargantuan construction project was almost at an end. In just a few short months, the last mighty slab of CMS would be hoisted by the giant crane attached to the roof of the hangar and slowly and delicately lowered into position.

My small role in all this was to write some computer code to convert the amount of light detected by the sensors that were glued to the back of each of these crystal blocks into a measurement of energy. Before I saw CMS, it had seemed like a pretty unstimulating task, but after that visit, I realized how fortunate I was to be able to contribute in even a minor way to such a Herculean effort.

There was a palpable tension in the air over those hot summer months. As we students got on with our little projects, sought out what scant nightlife was available in Geneva on a student budget, and sat bleary-eyed in lectures the next morning, thousands of physicists, engineers, and computer scientists were working determinedly to prepare for the start of the greatest experiment ever attempted by the human race.

The first major target of this colossal machine would be the Higgs boson, a particle originally predicted almost half a century earlier and one that is the keystone of our modern understanding of the make-up of matter. It is also the last known ingredient that we've yet to include in our apple pie recipe. To understand why the Higgs boson is so important and why so many decades of work and billions of dollars were spent in finding it, we are going to need to go even deeper into the strange world of quantum fields. A health warning: what follows includes some truly mind-melting stuff, but, if you persevere, I will try to show you, as best I can, some of the most profound and beautiful ideas in modern science.

HIDDEN SYMMETRY, UNIFICATION, AND THE BIRTH OF A BOSON

To understand why the Higgs is so darn important, we need to go right back to basics and think about the structure of matter. Everything is made of atoms, and an atom is made of negatively charged electrons orbiting a tiny positively charged nucleus. Probe deep inside the nucleus and you find protons and neutrons, go deeper still and you discover that protons and neutrons are made of up and down quarks. Thus, all matter is made from just three fundamental particles: electrons, up quarks, and down quarks. So far so familiar.

Of course, matter is not just the sum of its parts. The structure of an atom is as much determined by the forces that hold it together as it is by its building blocks. We have already encountered two of these forces: the electromagnetic force that binds electrons to the nucleus, and the strong force that holds quarks together inside protons and neutrons. Both of these forces are communicated by quantum fields – the electromagnetic field and the gluon fields – and if you put a discrete packet of energy into one of these quantum fields you create a little quantized ripple (aka a particle), known as a photon or a gluon.

However, there is a third force that we haven't discussed in detail, arguably the weirdest of all the fundamental forces: the weak force. The weak force is unique among the known forces as the only one that allows one type of fundamental particle to transform into another. The first evidence for the weak force's influence was Henri Becquerel's discovery of radioactivity back in 1896. As Ernest Rutherford described a couple of years later, one type of radioactivity known as beta decay involved an unstable nucleus spitting out an electron. This led to no end of confusion for many years, as physicists not unreasonably assumed that if an electron came out of a nucleus, then it must have existed in the nucleus to start with. However, by the 1930s it was realized that this was wrong. There are no electrons in the nucleus; instead during beta decay a neutron *transforms* into a proton, an electron, and an anti-neutrino, and it does this via a new fundamental force, the weak force.*

However, in the 1930s the true nature of the weak force was far from understood. A successful theory had been written down by the Italian-American superstar Enrico Fermi, who described beta decay in terms of a neutron decaying directly into a proton, electron, and antineutrino without the need for any extra force fields to help the decay along. It became clear though that Fermi's theory was only an approximate description. When you calculated its consequences at higher and higher energies eventually the theory broke down and gave nonsensical answers, spitting out probabilities that were bigger than 100 per cent.

Clearly, a more fundamental theory was needed, and by the

---

* We now know that neutrons and protons are made of quarks, up-down-down in the case of a neutron, up-up-down in the case of a proton, so at the fundamental level what happens is that a down quark turns into an up quark, an electron, and an antineutrino.

1950s physicists had found what seemed like the perfect candidate: quantum field theory. The first successful quantum field theory described the electromagnetic force. Known as 'quantum electrodynamics', it was assembled over a period of years by a stellar cast of physicists including Hans Bethe, Freeman Dyson, Richard Feynman, Julian Schwinger, and Shin'ichirō Tomonaga. Quantum electrodynamics, or QED for short, is the most precise scientific theory ever written down, describing the way charged particles interact with the electromagnetic field with dazzling precision, in some cases making predictions that agree with experiments to better than 1 part in 10 billion.

At the heart of this stonkingly successful theory was a truly beautiful principle, the principle of 'local gauge symmetry'. First introduced by Julian Schwinger, this principle implied something utterly magical – that the fundamental forces arise because of deep symmetries in the laws of nature.

That's quite a claim, and it requires a bit of unpacking. First, we need to take one step back and consider the wider role of symmetry in physics. The first person to truly understand the power of symmetry in shaping the physical world was the brilliant German mathematician Emmy Noether. Her greatest gift to physics was Noether's theorem, which shows that if the universe is symmetric in a particular way, then there must be a corresponding quantity that is always conserved. What do we mean by symmetric? Well, as we've already seen, a symmetry exists if there is something you can do to a system, be it a physical object or the entire universe, that leaves it unchanged. Take a square and turn it by 90, 180, 270, or 360 degrees and it will look exactly the same as before you rotated it – a square has rotational symmetry.

The same is true for the laws of nature themselves. Imagine that you are a scientist on board an interstellar spaceship, way out in deep space, far beyond the gravitational influence of the Earth or the Sun. You could ask a question: does the direction that my spaceship is pointing make any difference to the results of any experiment I could think of doing on board? If the answer to this question is that it makes no difference, then we can say that the laws of nature are symmetric under rotations in space. Or put another way, the universe doesn't care which way you are pointing.*

Noether's theorem tells us that this rotational symmetry implies the existence of a conserved quantity – one that never gets bigger or smaller – which turns out to be angular momentum, the amount of rotational oomph possessed by a system.† Countless experiments have shown that it is always conserved; it cannot increase or decrease, it can only be redistributed among the component parts of a system. This means that if you set a rigid object spinning in the vacuum of space, it will keep on spinning forever at exactly the same rate, so long as nothing comes along and bumps into it. This is why Earth keeps turning so reliably – day always follows night because of the rotational symmetry of the laws of nature.

The same symmetry principle explains why energy and momentum are always conserved. Energy conservation is due to the fact that the laws of nature don't change with time, while the conservation of momentum is due to the fact that the laws of nature are the same everywhere in space. However, there is an even more remarkable consequence of symmetry: symmetries seem to be responsible for the forces of nature.

The symmetries connected to the conservation laws that I just described are called 'global symmetries', which means they involve transformations that are the same at every point in time and space. For instance, a global transformation could involve rotating the

* Of course, this wouldn't be true if you were close to a massive body like the Earth. The gravity of the Earth defines a preferred direction in space, which breaks the rotational symmetry.
† Non-quantum mechanically speaking, angular momentum depends upon the size, shape, and mass of an object and how fast it's rotating. Subatomic particles can also possess angular momentum in the form of spin.

whole universe by 90 degrees or shifting the entire universe 1 metre to the right. If the universe looks the same after you've done this, then it possesses a global symmetry.

However, the symmetries connected to the fundamental forces are *local* symmetries, meaning that they involve transformations that vary in time and space. The sorts of local transformations that play a role in particle physics are rather harder to visualize, so let's start with an analogy.

Consider two teams playing a game of football on a flat, well-manicured field. Now imagine that we had the godlike power to raise the level of the field by an arbitrary amount, it could be a metre or a kilometre. Apart from the fact that the players might struggle to breathe if we lifted the field too high, as long as we raised the whole field by the same amount, this should have absolutely no effect on how the game is played. As a result, we can say that football is symmetric under a global change of field height.

Now let's imagine that instead of raising the entire field by the same amount, we somehow conspired to tip it at an angle so that one team found themselves playing uphill and the other downhill. This would count as a local transformation since the change of ground level depends on where you are on the field. Clearly, such a transformation would have a big impact on the game; I imagine (though as my friends will tell you I know absolutely f-all about football) it would be far easier to reach the goal at the bottom of the field than the one at the top, giving one team a big advantage. In other words, this local transformation does not result in a symmetry.

However, what if we were being really stubborn and insisted that we somehow wanted to restore fairness to the game? Well, one way to do this would be to use our aforementioned godlike powers to create a wind that blew constantly uphill, making it harder for the team playing downhill to reach the goal and easier for the team playing uphill. In other words, symmetry is restored by introducing a force.

Amazingly, this is not a million miles from how the electromagnetic force arises in QED, except that instead of a game of football we are now considering the rules governing the way electrons and other charged particles behave. As we've discussed over the last couple of chapters, an electron is a wave or vibration in the electron field. Just like a wave on the ocean, this wave changes with time. If you look at one point in space, sometimes the vibration in the electron field will be big, at other times it will be small, following a characteristic cycle as it wobbles up and down. Where you are in this cycle is known as the 'phase' of the wave. You can think of the phase as like a little clock that tells you how far through a wobble in the electron field you are.

Just like with the level of our imaginary football field, it is possible to ask what happens if we perform transformations to the electron field's phase. First, let's say we make a global transformation, shifting the electron's internal clock by a uniform amount, say a half-cycle, everywhere in space and time. According to Noether's theorem, if this global transformation has no effect on how the electron field behaves, then a conserved quantity exists, which, rather incredibly, turns out to be none other than electric charge. Or to put it another way, electric charge is conserved because of a global symmetry.

Now for the really amazing part. Let's say we introduce a phase shift that varies in space and time. Imagine, if you will, a vast array of tiny clocks describing the phase of the electron field, one for each point in space and time. A local transformation could mean that over here the hand of the clock is shifted forwards a quarter-turn, while over there it's shifted back by a half. If we want to be able to introduce different phase shifts at different places and times *without* affecting the way the electron field behaves then we discover that we have no choice but to introduce a new quantum field. And most remarkably of all, this new field has the precise properties of the electromagnetic field. The electromagnetic field acts like the wind on our sloped football field, correcting the effect of shifting the phase of the electron field unevenly through space and time.

It's worth pausing for a moment to appreciate how tremendous an insight this is. According to QED, your fridge magnet sticks to your fridge, electric currents flow through wires, and atoms have the structure that they do ultimately because of deep symmetries in the laws of nature. When I first learned this fact as an undergraduate, it left me in awe. Years later it still feels a little bit like magic.

In QED, the collection of phase transformations that gives rise to the electromagnetic field is described by a mathematical object known as the U(1) symmetry group. What this group is in detail doesn't really matter for our purposes, but the important point is that if you demand that the laws of nature don't change when you perform a local U(1) phase transformation, then the electromagnetic field must exist. Even better, the mathematical structure of U(1) completely determines the rules of electromagnetism and all the phenomena that depend on them, from the way sunlight reflects off the surface of a lake to the awesome power of a lightning storm. Importantly, it also means that the photon, the particle of light, has to be massless. The equivalent in our football analogy would be to discover a deep symmetry principle that automatically generates the complete rulebook for the game of football, including the offside rule and specifying the size of the ball.

This idea of symmetry now leads us back to where we started: the Higgs and the problem of the weak force. Once QED was discovered, physicists naturally tried to see if other quantum field theories describing the weak and strong force could be found based on similar symmetry principles. A class of such theories was discovered by Chen Ning Yang and Robert Mills in 1954, but they all suffered from what seemed to be a terminal problem – they predicted the existence of new massless particles. These particles would be similar to photons – the particles of the electromagnetic field – in that they would have no mass, but different in that they carried a charge. The problem was that if such particles existed then they ought to be flying about all over the place, meaning they should have been discovered long ago. As a result, most physicists believed that so-called Yang–Mills theories, based on similar symmetry principles to electromagnetism, were nonstarters.

Now, as we already saw in the case of the strong force, such massless particles *do* exist – they're called gluons – but they hadn't been discovered in 1954 because gluons are inexorably locked inside protons and neutrons, thanks to the tremendous strength of the strong force.

However, the absence of massless particles can't be explained away like this for the weak force. The symmetry group that was identified as being the most promising candidate for a quantum field theory of the weak force is known as SU(2), and it predicts three new force fields, along with their corresponding massless particles: the W^+, W^-, and the Z^0 bosons.

In case you were wondering what this boson business is about, a brief aside. Particles are divided into two categories depending on their spins. As we've seen, quantum mechanical spin comes in lumps of ½ and particles with so-called half-integer spins – those in the sequence ½, $\frac{3}{2}$, $\frac{5}{2}$, etc. – are known as fermions. The matter particles, including electrons and quarks, are all spin ½ fermions. Bosons, on the other hand, have integer spins in the sequence 0, 1, 2, etc. and include the force particles like the photon, gluons, and W and Z bosons, which all have spin 1.

This SU(2) theory had lots of attractive features; for instance, it could explain what was really going on in beta decay. Instead of a neutron decaying directly into a proton, electron, and antineutrino all in one go, as in Fermi's theory, the neutron now decayed into a proton by giving off a W^- boson, which then converted into an electron and an antineutrino.

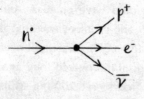

Fermi theory: Neutron (n⁰) turns straight into a proton (p⁺), electron (e⁻), and an antineutrino.

SU(2) theory: Neutron (n⁰) turns into a proton (p⁺) and a W⁻ boson, which then turns into an electron (e⁻) and an antineutrino ($\bar{v}$).

However, there is a big problem with this theory. Since it predicts that the W^+, W^-, and Z^0 bosons should be massless, it should be incredibly easy to make them, for instance when two particles bang into each other. So just like light, which is made up of photons, we ought to see them flying about all over the place in the real world. The fact that no one had ever seen a W or Z boson meant that the theory had to be wrong. Worse still, massless particles would turn the weak force into a strong force.

For force fields like the electromagnetic field or the weak fields, the mass of the particle that goes with the field can be thought of as a kind of energy toll for using that field, a bit like the toll you might pay for crossing a bridge. Since photons have zero mass, the cost of crossing the electromagnetic field is zero, which means two things. First, if you have two electrically charged particles, they will exert a force on each other via the electromagnetic field no matter how far apart they are. Of course, the force gets weaker the bigger the distance between them, but nonetheless, a force is still exerted. In physics parlance we therefore say that the electromagnetic force is 'long ranged'. The second consequence of the fact that the toll for using the electromagnetic field is zero is that the electromagnetic force is a relatively strong force.

If the weak particles were also massless the same would be true of the weak force – it would be long ranged and strong. However, as the name suggests, the weak force is, well, weak and extremely short ranged; its effects only become apparent at distances smaller than an atomic nucleus, which is why we don't notice the weak force in our everyday lives.

One way to keep the weak force weak would be to give the W and Z particles very large masses – in other words, introduce a large energy toll for using the weak fields. This would be like charging £1,000 for every metre you drive across a bridge, which would mean only the richest drivers (that is, those with the most energy) would use it, and the journeys would tend to be very short. Consequently, large masses would make the weak force weak and short ranged and also neatly explain why no one had seen a W or a Z particle in the 1950s and 1960s – if their masses were sufficiently large, they would be too heavy to have been created in any experiment at the time.

However, there was a serious problem with this fix. Giving mass to the W and Z particles broke the beautiful SU(2) symmetry that was used to determine the form of the weak force in the first place. Even worse, the theory became plagued by infinities – calculations of probabilities that gave infinite answers – rendering it effectively useless.

Theorists had run into similar infinities when QED was being put together during the 1930s and 1940s, but these had eventually been overcome, turning QED into a theory that gave sensible answers. This technique, known as 'renormalization', was so crucial to the success of QED that it won Schwinger, Feynman, and Tomonaga the 1965 Nobel Prize. However, renormalization didn't seem to work for a weak force that assumed massive particles. The way to a quantum field theory of the weak force seemed well and truly blocked.

It would take a decade for the solution to be found, and yet more time for that solution to be incorporated into a fully functioning theory of the real world. The story of its discovery is long and tangled and was a group effort featuring many star players. Julian Schwinger, Yoichiro Nambu, Jeffrey Goldstone, and Philip Anderson laid the foundations; Robert Brout, François Englert, Peter Higgs, Gerald Guralnik, Carl Hagen, and Tom Kibble hit upon a potential solution; Abdus Salam, Sheldon Glashow, and Steven Weinberg applied this solution to the weak force; and, finally, Gerard 't Hooft and Martin Veltman proved the resulting theory was free of infinities. This complex history behind the theory could fill a book in its own right,* so instead I will simply focus on the physics. The final theory is like a cathedral, towering, beautiful, the work of many hands over many years. It is the heart of the standard model of particle physics.

Physicists were facing a paradox: they knew that the particles of the weak force must have large masses, or else the weak force would be strong and long ranged like electromagnetism, which it isn't. However, massive particles resulted in a theory that gave infinite results and destroyed the jewel-like symmetry that determined the form of the weak force in the first place.

But what if the symmetry wasn't destroyed? What if it was just hidden? Or to put it another way, what if the weak particles were *fundamentally* massless but got their masses from somewhere else? Enter the Higgs field.

The Higgs field is a brand-new type of quantum field that's unlike anything that we've seen so far. All the fields we've met are either matter fields with spin ½, like the electron, or spin 1 force fields, like the photon. This new field, uniquely, needed to have a spin of 0.

It is also unique in another crucial respect: it needed to have a nonzero value everywhere in space. This is very different from, say, the electromagnetic field. If you go to a really empty bit of space and remove all the photons so that there are no ripples sloshing about in the electromagnetic field, then the value of the field is pretty much zero, apart from some gentle jittering due to quantum uncertainty. However, even when you remove all the particles from

---

* If you'd like to know more of the history of who did what, and who may or may not deserve the credit, I can highly recommend *The Infinity Puzzle* by Frank Close.

the Higgs field, it still has a large nonzero value, effectively filling the entire universe with a uniform Higgs field soup.

The key insight was that such a Higgs soup could give masses to the weak particles. The story goes something like this. Way, way back in the very first moments of the universe's history the Higgs field's value was zero and as a result the three weak particles, the W^+, W^-, and Z^0, all had zero mass and symmetry reigned. However, after around a trillionth of a second, the Higgs field 'switched on', going from zero to a fixed value, filling the universe with a Higgs field soup and making the weak particles suddenly appear massive. When this happened, the perfect symmetry that was manifest at the beginning of time became hidden, and what we think of today as the weak force changed: from strong and long ranged to weak and short ranged.

At the same time, the particles of matter that would go on to make our apple pie, including electrons and quarks, which had previously been zipping through the universe at the speed of light, suddenly found themselves ploughing through this thick Higgs soup. As they interacted with the Higgs field, they too transformed from zippy massless particles into ponderous massive ones. If it helps, an imperfect analogy is to imagine the Higgs field as a gloopy substance that sticks to particles like electrons and quarks, slowing them down and imbuing them with the property of mass. Meanwhile, particles like photons and gluons remain massless because they don't interact directly with the Higgs field.

So the Higgs field isn't just responsible for giving mass to the particles of the weak force, it gives mass to the fundamental matter particles as well.* This makes it an absolutely essential ingredient

---

* One important qualification of all this is that the Higgs field only gives mass to fundamental matter particles, including electrons and quarks. However, most of the mass of protons and neutrons does not come from their constituent quarks but from the energy stored in the gluon fields that bind the quarks together. That means that most of the mass of an atom actually comes from the strong force, not the Higgs.

of our apple pie and, by extension, our universe. Without the Higgs field, the world as we know it could not exist. Particles like electrons would have no mass, meaning that they would zip around at the speed of light and never bind to make atoms. At the same time the forces of nature that we know and love would be utterly transformed. Precisely what this Higgsless universe would look like is hard to say, but it would certainly not be a place where we could live.

The basic principles of this mechanism were first published in 1964 by three independent groups: first into print were Robert Brout and François Englert in Brussels, next Peter Higgs in Edinburgh, and finally Gerald Guralnik, Carl Hagen, and Tom Kibble in London. Why then, you might ask, is Peter Higgs the only one whose name got attached to the idea? Well, in essence it's down to an unfair quirk of history. Higgs himself, who is unfailingly self-effacing, refers to it as the ABEGHHK'tH* mechanism, in credit to the many theorists who contributed to the idea. Sadly, saying that out loud tends to make people think that you're trying to regurgitate a hairball.

There is one thing that marked Higgs out from the rest of the pack. When the first draft of his paper was rejected by the journal, Higgs decided to beef it up by adding some experimental consequences of his idea. Inventing a new cosmic energy field is all well and good, but how do you know if such a thing exists? Higgs knew that like all other quantum fields it should be possible to create a ripple in this new field, which would show up in experiments as a new particle. Now, to anyone with training in quantum field theory, the fact that a particle should come along with the Higgs field was pretty obvious, but since no one else had thought to mention it explicitly, Higgs's name became forever associated with that particle – the famous Higgs boson.

---

* That stands for Anderson, Brout, Englert, Guralnik, Hagen, Higgs, Kibble, 't Hooft.

What the gang of six had outlined was the basic principle of giving mass to particles using a new quantum field. However, it had yet to be fully adapted to describe the weak force. In 1968, Sheldon Glashow, Abdus Salam, and Steven Weinberg used it to create a fully consistent theory of the real world. In doing so, they discovered that the only way to make the mass-giving mechanism work was to include the electromagnetic force in the theory as well, which led to one of the most profound discoveries of the twentieth century: the electromagnetic and weak forces, which appear utterly different in our ordinary world, are in truth different aspects of one unified *electroweak* force. The only reason we see two separate forces today is that early in the universe, the Higgs field gave mass to the W and Z bosons but left the photon massless.

This truly incredible revelation marked the greatest unification in physics since Faraday and Maxwell had shown that electricity and magnetism were one and the same phenomenon, aspects of a single, unified electromagnetic field. Now the electromagnetic field had been united with the weak force through the application of deep symmetry principles. The resulting electroweak theory predicted the existence of three new massive force particles, the W^+, W^-, and Z^0 bosons, which were spectacularly discovered at CERN's Super Proton Synchrotron collider in 1983. At long last, the weak force was understood and the electroweak theory became the core of the modern standard model of particle physics.

However, one piece of the puzzle had yet to be found: the Higgs boson itself. Without the Higgs, the cornerstone of the beautiful theoretical cathedral built in the 1960s and 1970s was missing. It was the Higgs that explained the strength of the weak force, unified it with electromagnetism, and gave mass to the particles that make up every atom in the universe. That is why finding it was so crucial, and that is why, in the late 1970s, some farsighted physicists at CERN began to plan the most audacious scientific experiment ever attempted.

THE BIG BANG MACHINE

On Tuesday, 30 March 2010, just before lunchtime, two unsuspecting protons are about to make history. Their day began ordinarily enough, bouncing around contentedly in a canister of hydrogen gas in a nondescript surface building at CERN, a few miles from downtown Geneva. It is a day that should have been utterly forgettable. After all, they have both lived unimaginably long and varied lives by human standards, born as they were 13.8 billion years ago in the furious heat of the big bang.

The things they've seen! They witnessed the searing light of creation and the endless dark before the first stars. They danced in the shining atmospheres of blue hypergiants and surfed the cosmos on the shockwaves of supernovae. One of them had even spent some time in a skin cell on the tip of Paul McCartney's left middle finger, albeit during his time in Wings.

However, unbeknownst to them, their long lives are about to be cut brutally short. But for the misfortune of ending up in that particular bottle of hydrogen gas, they might have lived to see the passing of the human race, the death of the Sun, and perhaps even the coming of the second great darkness at the end of the universe. Unfortunately, that particular canister has been chosen to act as the proton source for the most powerful particle collider on planet Earth. They are about to become martyrs to the cause of science.

Without warning, the opening of a valve sucks them unceremoniously from their gas bottle into an adjacent metal box, where a fierce jolt of electrical energy rips them rudely from their hydrogen molecules. Bidding farewell to their companion electrons, the protons find themselves naked and alone, hurtling down an evacuated pipe running through the centre of the first accelerator in a long chain that will lead them, inexorably, to their doom. When they exit the prosaically named Linear Accelerator 2 after a short trip of just 30 metres, they are already travelling at a third of the speed of light. Neither of our two protons has travelled this fast for billions of years, but still, a third of the speed of light is nothing to get too worried about. Many of their companions rained down on the Earth as cosmic rays at far higher speeds.

However, as they pass through a series of increasingly large circular accelerators, their alarm begins to grow. With each turn, powerful electric fields give them a kick in energy, while stronger and stronger magnetic fields hold them more and more tightly on their orbits. Before long they find themselves racing around a huge 7-kilometre-circumference ring, the Super Proton Synchrotron (SPS), once CERN's mightiest accelerator. Each proton is now flying in a tightly packed swarm with billions of its former canistermates as they are gradually accelerated to ever-more terrifying speeds. When the SPS maxes out at 99.9998 per cent of the speed of light, they hope their ordeal might nearly be over.

It is not. A sudden magnetic jolt kicks the protons out of the SPS and along a transfer line leading to an even larger ring, the largest of all in fact, the Large Hadron Collider. Ominously, our two protons now find themselves moving around the 27-kilometre ring in *opposite* directions. That can't be good. They get some brief comfort from the fact that the gentler curve of the larger machine means they are being yanked on less fiercely by the LHC's magnets. Alas it is not to last.

At 11.40 a.m., with two fully loaded beams circling in opposite directions, the LHC begins to push the protons into uncharted territory. On each orbit of the ring, they pass through a short stretch of metal cavities where they are subjected to a cascade of violent smacks from a 2-million-volt electric field, pushing them ever closer to the speed of light. At the same time, more than a thousand superconducting magnets that make up the bulk of the collider begin to generate increasingly powerful magnetic fields, pulling the protons towards the centre of the ring with increasingly terrible force.

For an hour the protons endure this electromagnetic torture. Mercifully, thanks to their tremendous speeds and the time-warping effects of relativity, this hour passes in little over a second from the protons' point of view. At 12.38 p.m., after around 40 million orbits of the 27-kilometre ring, they reach their final, dizzying speed: 99.999996 per cent of the speed of light.

Each proton has been transformed into a projectile of unrivalled potency, carrying 3.5 trillion electron volts (TeV) of energy, approximately 3,700 times its own rest mass. As they flash around the ring once every 90 microseconds, they repeatedly pass through the hearts of each of the four vast detectors – ATLAS, ALICE, CMS, and LHCb – which wait patiently to record their demise. For now, the two oppositely rotating beams are kept apart, coming within a few millimetres each time they pass through the detectors, but not yet meeting.

A few short minutes later, at 12.56 p.m., the protons find themselves slowly shifting in their orbits as sets of magnets positioned on either side of the four detectors begin to guide the two counter-rotating beams closer and closer together. With each orbit, the gap between them narrows, until there is less than a whisker between them.

12.58 p.m. Our two protons, now moving faster than they have since the beginning of the universe, pass each other one last time and begin their final approach from a distance of 27 kilometres. A mere 22.5 microseconds later the distance between them has halved. Another 22.5 microseconds and they enter one of the detector caverns from opposite ends. The only mercy is that they can't see each other coming, being protons and not having eyes, and also because the inside of the LHC is pitch black.

As the two mighty beams of the LHC cross paths inside all four detectors, our hapless protons run headlong into each other with a violence not seen since the big bang. The world record for the highest energy collision ever produced by science is smashed.

The force of the impact utterly obliterates the protons, spraying their innards outwards in a firework of quarks and gluons. At the same time, their vast energy sends ripples cascading through the quantum fields of nature, creating new particles from the force of the impact: electrons, muons, photons, gluons, quarks, and many more besides. As our two protons die, new particles are born. Their deaths make new matter from energy. $E = mc^2$.

A hundred metres aboveground, as images of the first collisions flash up on screens in control rooms around the LHC ring, engineers and physicists erupt in jubilation. After more than thirty years of dreaming, planning, construction, testing, setbacks, and recoveries, the most ambitious scientific experiment ever conceived has finally begun. The achievement is all the sweeter for the engineers in charge of the LHC in the CERN Control Centre, many of whom have spent the last year working tirelessly down in the 27-kilometre tunnel repairing the collider after it literally blew itself apart shortly after it first switched on in September 2008. As champagne bottles are uncorked across CERN, Mirko Pojer, one of the two engineers operating the LHC that day, simply notes in the log, 'First collisions at 3.5 TeV per beam!'

Tuesday, 30 March 2010, was a big day for everyone at CERN. It marked the start of the LHC's physics programme and with it the beginning of the search for answers to some of the deepest questions it is possible to ask about our universe. I still remember those first few weeks vividly: the excitement of seeing the first event displays showing particles flying through the LHCb detector, the sense of responsibility and pressure that came with helping to keep one small part of the experiment running smoothly, the thrill of seeing the first data showing that *real* particles were being made in *our* detector.

As the days and weeks passed, each of the four LHC experiments recorded more and more collisions at an ever-increasing rate, as the engineers driving the LHC from the CERN Control Centre gradually learned how to operate the impossibly complex machine that they had built. Having set a new world record for the highest energy collisions ever created in a lab, their task was now to figure out how to increase the *rate* of collisions, so that more and more data could be recorded each day by the four experiments.

LHCb, the experiment I work on, is a specialized detector designed primarily to study particles known as 'bottom quarks'. These are heavy companions of the ordinary down quarks that make up protons and neutrons, and by measuring their properties in detail we can learn a great deal about both the standard model and potentially new quantum fields that we haven't yet discovered. However, one thing LHCb is not designed to do is to search for the Higgs boson.

The Higgs was the target of the two largest experiments at the LHC: ATLAS and CMS.

These two behemoths are 'general-purpose' detectors, which means that they're designed to search for as wide a variety of different particles as possible. They sit on opposite sides of the ring; ATLAS is just next to the main CERN site, and therefore close to all the handy amenities like the restaurants, while CMS is several kilometres away in the middle of the French countryside, which is very picturesque but must be a bit of a pain if you fancy popping back to CERN for lunch.

Each of these two experiments is staffed by a team of more than three thousand physicists, engineers, and computer scientists from across the world, each making a small, or sometimes not so small, contribution to the ultimate goal of discovering new features of the subatomic landscape. There are the senior physicists, many of whom were involved in the early planning of the experiments and now have managerial roles in the collaboration, setting strategy and leading the not inconsiderable task of building consensus among three thousand sometimes oversized egos. Then there are the hardware and software experts who designed and built the experimental equipment and software stack, who are absolutely critical to running and maintaining the experiment, while planning and delivering future upgrades. And then there is a veritable army of physicists, many of them young PhD students and postdoctoral researchers, whose job is to analyse the relentless flood of data and search for signs of new particles.

The experiments are only possible thanks to this enormous international group effort, and yet remarkably, the search for the Higgs boson ultimately came down to just a handful of young people who had the awesome responsibility of bringing a fifty-year search to its dramatic conclusion and, with it, providing the answer to the question, why do particles have mass? One of these fortunate few was Matt Kenzie, a former colleague of mine at Cambridge who back in those early days was a PhD student at Imperial College London just starting out on CMS.

Matt first arrived at CERN in the spring of 2011, just as the LHC was firing up for its second year of collisions. He had started his physics career with aspirations of being a theorist and perhaps doing something small like discovering a quantum theory of gravity. However, after a year studying for his master's, he realized the life of a theorist wasn't for him and ended up submitting a last-minute application for a PhD in experimental particle physics. So last-minute in fact that the deadline for applications had already passed, but by a stroke of luck Imperial had a spare place going after one of the candidates they'd made an offer to had dropped out. After a hastily arranged interview he soon found himself embarking on a PhD, and before he knew it, he was heading to CERN to work on the search for the Higgs boson.

Of the three thousand people on CMS, hundreds were involved in the Higgs search to greater or lesser degrees, but in reality, almost all of the day-to-day work was done by a relatively small group of a few dozen researchers. By Matt's own admission it was pure luck that he ended up as one of the select few who had the incredible privilege of analysing the data that would settle one of the longest-running questions in physics. 'It was really a complete fluke that I landed in the middle of this huge scoop,' he told me.

Physicists at ATLAS and CMS knew that if Higgs bosons got

created in the collisions at the LHC then they would decay almost instantaneously, within just 10^{-22} seconds, far too short a time for them to reach any of the sensitive parts of the detectors. That meant that the only way to find evidence for the Higgs would be to try to catch the particles it decayed into as they zipped out through the detector, a bit like trying to figure out the make and model of a car by filling it with dynamite and photographing the various bits of shrapnel as they fly past.

However, unlike a car, which always falls apart into the same basic bits, the Higgs can decay in a variety of different ways. Some of these are easier to spot than others. For example, its decay into a bottom quark and an anti-bottom quark is by far the most common, with more than half of all Higgs bosons decaying this way, but nonetheless it is incredibly hard to spot thanks to the vast numbers of bottom quarks that get produced by the LHC collisions. Searching for the Higgs decaying to a bottom–anti-bottom pair is less like looking for a needle in a haystack than looking for a needle in a pile of other very similar-looking needles.

Luckily, there are far better ways to spy a Higgs. The most promising decays of all are when a Higgs turns into either two high-energy photons or two Z bosons. That doesn't mean seeing these decays is easy, but it is at least more like looking for a needle in a haystack (or perhaps a needle in a field full of haystacks).

Matt's task on arriving at CERN was to help write the computer code that would search for signs of a Higgs boson decaying into two photons. He remembers getting a bit of a shock the first time he presented his work to one of CMS's working groups, receiving a particularly hostile roasting from a postdoctoral researcher from MIT. It turned out that there was quite a bit of rivalry between the MIT and Imperial groups, who were both vying to lead the search, and Matt had unwittingly found himself in the middle of this battle. 'It was a bit of a baptism of fire,' as he put it. 'I came away feeling this is quite a tough environment.'

At times the pressure was intense. Not only was Matt's team, which included physicists from Imperial, San Diego, CERN, and Italy, competing against the MIT group in the search for the Higgs decay to two photons, there were other groups at CMS working on the Higgs decay into two Z bosons, not to mention three thousand other researchers on their giant rival ATLAS, who were racing towards the same goal. Matt worked in a tight-knit team with a couple of other PhD students and two postdoctoral researchers. They'd meet at least twice a week, often talking through their latest results over breakfast or lunch in CERN's Restaurant 1, the bustling cafeteria that is the hub of social activity at the lab. There were a lot of long evenings working in the office after hours, and one or two all-nighters, particularly in the lead up to public updates on the Higgs hunt.

So how do you actually hunt for the Higgs? Well, it all begins with the collisions produced by the LHC. Let's say that the Higgs field exists, and that when protons smash into each other at the LHC they have enough energy to set the Higgs field wobbling, creating a new particle, the Higgs boson. The first challenge is that the chances of producing a Higgs in a single collision are extremely low. Thanks to the probabilistic nature of quantum mechanics, when two protons collide you cannot say ahead of time what particles will be produced. A collision is like rolling a die with an enormous number of faces, with each face corresponding to a different potential outcome.

Most of the time a collision will just produce particles we already know about: quarks, gluons, photons, maybe a W or Z boson. The odds of making a Higgs are fantastically slim – just two in a billion – so to have any chance of seeing a decent number the LHC had to create stupendously large numbers of collisions. In fact, it is able to produce around a billion collisions every second inside ATLAS and CMS, and the aim is to do this twenty-four hours a day, seven days a week, for about nine months of the year, minus a few technical stops and the time it takes to set up for a new run. By the end of 2012 this incredible collision rate meant the LHC had produced six quadrillion (thousand trillion) collisions.

This vast number of collisions implies a corresponding vast amount of data, so much, in fact, that the LHC could fill every hard drive on Earth in a matter of days. The only way to cope with this digital tsunami is to throw away most of the collisions before they are even recorded. This is done by a set of very fast computer algorithms called 'triggers', which look at each collision in real time – once every 25 nanoseconds – and decide whether it looks like something interesting may have happened or whether it's just a load of boring old quarks and gluons that we've seen before. On the rare occasions it does look interesting, the data is recorded, and analysts like Matt can start their search.

Once the data is stored, the way you search for Higgs bosons that have decayed into two photons is actually rather straightforward, at least in principle. The teams at CMS and ATLAS used tailor-made algorithms to sift through their huge datasets in search of collisions containing a pair of high-energy photons. However, the hard part is that there is an enormous number of other ways that photons can get made when two protons smash into each other, and most of them don't involve a Higgs boson at all. Matt and his colleagues needed to find a way of sifting out the real Higgs bosons from this background of random photons that didn't come from a Higgs, a bit like panning for gold in a fast-flowing stream.

Even after the data is sifted, there will still be a lot more background than real Higgses, but here one final trick comes into play. If you add up the energies of the two photons, that directly tells you the mass of the particle they came from. For random background photons, their energies can add up to more or less any number, which means that if you plot them on a graph they will be spread out over a wide range of masses. However, photons that came from a Higgs will always add up to give the *same* mass – the mass of the Higgs boson – which means that they cluster around the same value, creating a little bump on top of the flat background. If this bump is prominent enough, then that is the smoking gun of a new particle being produced in your experiment.

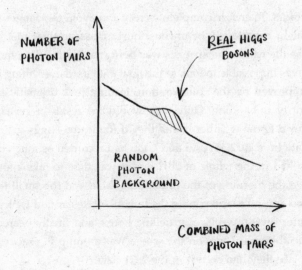

Just before Christmas in 2011, rumours began to swirl around the corridors of CERN that ATLAS and CMS had spotted something. After a year of intense data collecting and analysis, a special seminar was arranged on 13 December to update the scientific community on the search for the most elusive particle in history. Hundreds of people queued to get a seat in CERN's main auditorium to hear the presentations, forcing Matt to watch the proceedings via live webcast from his office. The results were tantalizing but inconclusive. Both ATLAS and CMS had seen a small bump at a mass of 125 GeV (for reference, a proton has a mass of around 1 GeV and a Z boson has a mass of 90 GeV). But it was too small to be sure that it wasn't just a statistical fluctuation. However, small as the bump might have been, the fact that two independent experiments had both seen hints of a new particle at the same mass set many pulses racing in the particle physics community.

When data collection began again in the spring of 2012, everything suddenly became much more serious. Sensing that both they and ATLAS were hot on the trail of the Higgs, Matt and the team had to work even harder to make sure they kept up with their competitors. To ensure complete secrecy and avoid the danger that they might consciously or unconsciously massage their methods and bias the result, the analysis was performed blind. This meant that it was impossible to look at the final data until everything had been approved by the collaboration, leading to a dramatic final moment of unblinding. Only then would the result be revealed, and they'd know whether or not they'd found the Higgs.

By late June 2012, ATLAS and CMS had recorded as much data as they had in the whole of 2011 and it was time to take another look. On the morning of the unblinding, Matt and the small team gathered around a laptop amid the hubbub of Restaurant 1, bleary-eyed after several weeks of gruelling hours, and finally peered at the crucial graph. There on the screen was a bump in exactly the same place they had seen it in the 2011 data.

Being so new to particle physics, Matt didn't realize the full significance of what they'd seen until their results were presented to the whole CMS collaboration later that afternoon. Hundreds of physicists squeezed into the seminar room to listen to Mingming Yang, an energetic Chinese PhD student on the MIT team, present the results of the search for the Higgs boson decaying into two photons. She teased the audience by slowly revealing the results in stages, first 2011, then 2012. As she came to the last slide showing the combined result she gave a final rhetorical flourish: 'I hope you remember this moment for the rest of your lives.' A click later and there on the projector screen was a graph with a clear spike at 125 GeV. When the team searching for the Higgs boson's decay into two Z bosons revealed a spike in exactly the same place a few minutes later, Matt said, 'All hell broke loose.'

The next day he got an email from Joe Incandela, the spokes-person of CMS, inviting him and around fifty other physicists to help him prepare for a public announcement that had been sched-uled for 4 July 2012 (Higgsdependence Day as it jokingly became known at CERN). Joe had the awesome responsibility of revealing CMS's results to the world, alongside Fabiola Gianotti, who would be presenting the results on behalf of ATLAS. The gang of fifty spent several days locked away in that seminar room, discussing how best to present the results and polishing the final presentation.

While this was going on, Matt was flying back and forth to the United Kingdom each weekend to see his father, who was seriously ill. He did his best to explain to his parents what had been found at CERN, and while they understood it was a big deal, he got the sense they didn't fully grasp what this Higgs business was all about. 'Their response was sort of "That's lovely, dear." They had more important things to worry about at the time.'

The upshot of this was that Matt hadn't expected to be at CERN for the announcement and so had turned down the offer of a reserved seat in the main auditorium. However, as things turned out, he *was* back at CERN on the crucial day and managed to sneak in with a few of the other members of the CMS Higgs team.

The atmosphere in the room was electric, with one physicist describing it as like being at a football match. Now I'm not sure he'd ever been at a football match, but it was certainly very lively by particle physics standards. People had camped outside the auditorium overnight for a chance to get a seat, with hundreds having to be turned away. Meanwhile, in London, I was at a huge live webcast event near Parliament with a bunch of sciencey types, journalists, and members of government, all waiting for the presentations to start.

A couple of minutes before kick-off, the director general of CERN, Rolf Heuer, entered the auditorium flanked by Peter Higgs and François Englert, whose original research had set off this incredible chain of events almost half a century earlier. It was then that Matt realized he was involved in something big.

First came the presentation from CMS, as Joe Incandela revealed the same peaks that they'd all seen a couple of weeks earlier. Now everyone was waiting with bated breath to see if ATLAS had seen the same thing. When Fabiola Gianotti revealed a graph showing a bump at exactly the same mass as CMS, the room erupted.

The assembled physicists hollered and cheered the tremendous joint achievement. Peter Higgs, now well into his eighties, could be seen wiping away a tear. He later said that he had never expected the particle he had first predicted as a young man in 1964 to be discovered in his lifetime. It was only when CERN's director general made his final declaration at the end of the event that Matt realized just what he had been involved in: 'As a layman, I would now say, I think we have it.'

They had found the Higgs boson.

WARNING

Science Under Construction

The discovery of the Higgs marks a turning point in our quest to find the ultimate apple pie recipe. The Higgs completes the standard model of particle physics, a theory that has been staggeringly successful in describing the basic ingredients of our universe and the laws that govern their behaviour. But despite its success, the standard model cannot explain a critical detail – namely, where the matter in our apple pie *came from*. There must be more to the story.

The discovery of the Higgs at the Large Hadron Collider tells us about the physics that was going on in the universe around a trillionth of a second after the big bang. We now have fairly solid evidence that about this time the Higgs field switched on, giving mass to the fundamental particles and setting up the basic ingredients of the universe as we know them today. However, what happened before this moment is still shrouded in uncertainty.

How did the particles in our apple pie come to be? Why does the universe contain the quantum fields that it does? Are there ingredients that we're still missing? How did the universe begin? To answer these questions, we must now push back even further into that first trillionth of a second. It was in that short but crucial epoch that the matter of which we, and everything in the universe, are made came into being. Almost all the great questions of modern physics and cosmology turn on what happened in that first instant after the universe burst into existence.

So here I must give a disclaimer, like a tour guide taking you off the well-trodden paths into uncertain terrain. The farther we go from here, the less sure of our footing we become. We are entering a world of speculation, where sometimes even the questions aren't clearly formed, let alone the answers. But that's where Carl Sagan's challenge leads us. It's time to invent the universe.

CHAPTER 11

The Recipe for Everything

Around a millionth of a second after the big bang everything almost ended.

During the first microsecond, the universe had been so incredibly hot that particles and antiparticles were constantly being created and destroyed. Quarks and antiquarks, electrons and antielectrons flashed in and out of existence, emerging from the seething plasma in particle–antiparticle pairs only to annihilate again an instant later.

Meanwhile the universe had been rapidly expanding and cooling. Around a millionth of a second in, there was no longer enough heat in the plasma to create new protons and antiprotons and the apocalypse began. Particles and antiparticles destroyed each other in a great annihilation, wiping out almost all the matter in the universe in an almighty blast of radiation. This cataclysm should have spelled the end of all matter and antimatter, leaving a vast, dark, empty void with only a few lonely photons coasting through the endless nothingness.

But somehow, around 1 particle in 10 billion survived. We don't know how this happened. But it is only thanks to this 1 in 10 billion imbalance between matter and antimatter that the material universe – galaxies, stars, planets, human beings, apple pies – exists.

For all its success in describing the behaviour of the fundamental particles that make up our world, the standard model of particle physics predicts that the material universe should not exist. Now any theory that predicts the nonexistence of its own authors is in fairly serious trouble, which is one reason why physicists are convinced there must be something new still to discover.

The problem can be traced back to a time long before the standard model was first assembled, to 1928, when a young Paul Dirac saw antielectrons emerging from his famous equation. Even back then, Dirac knew that if antiparticles existed then they should always be created together with ordinary particles. Make an electron, said Dirac, and you must also make an antielectron. Every experiment performed since has proved Dirac right. It is true that the Large Hadron Collider makes matter out of energy, but add up all the particles that are created in a collision and you will always find an equal number of antiparticles. It seems to be impossible to make or destroy a particle without doing the same to an antiparticle.

This perfect balance between matter and antimatter should have led to an empty universe, and yet here we are. This is one of the biggest mysteries in modern physics, and attempts to explain it generally involve new, hitherto undiscovered quantum fields.

That said, it is possible to imagine a way around this problem that doesn't involve any new particle physics. What if, instead of particles totally annihilating one another, random motion in the churning primordial plasma randomly led to some regions where there was more matter and some where there was more antimatter? Fast-forward to the present day, and these regions would have been blown up by the expansion of space to cover vast tracts of the cosmos, some containing ordinary gas, dust, stars, and galaxies and others anti-gas, anti-dust, anti-stars, and anti-galaxies. From here on Earth, a distant anti-galaxy would look no different from an ordinary one, so perhaps some of the galaxies in the night sky are made of antimatter.

It's a neat idea; the trouble is that if there were really regions of the universe made of antimatter, then there would inevitably be boundaries where they pushed up against ordinary matter regions. Even the huge, empty spaces between galaxies contain small amounts of hydrogen and helium gas, so wherever such a boundary occurred you'd expect to see telltale gamma rays being produced by the annihilation of gas and anti-gas. The fact that we don't see any signals from such annihilations anywhere in the night sky suggests that the entire observable universe is made only from ordinary matter.

So the only explanation left to us is that sometime in the first moments of the universe's existence, something happened that allowed a tiny bit more matter to form than antimatter. That tiny imbalance – just 10 billion and 1 protons for every 10 billion anti-protons – allowed enough matter to survive the great annihilation to create everything we see around us today. However, finding a way to even create an imbalance this small proves to be incredibly difficult.

One of the first people to have a go was the Russian theoretical physicist Andrei Sakharov, who laid out three conditions that had to be satisfied in order for matter to be made in the early universe. They're known as the Sakharov conditions:

1. A process must exist that allows you to make more quarks than antiquarks.
2. The symmetry that relates matter to antimatter has to be imperfect.
3. When this matter-making process happened, the universe needed to be out of thermal equilibrium.

Condition 1 is probably the most straightforward to understand; clearly if we want to be able to make more matter than antimatter, we need a process that can do this. However, this isn't enough on its own, because even if such a process exists, then the symmetry between matter and antimatter would imply a mirror image process that makes more antimatter than matter. Hence, we need condition 2, which insists that the symmetry between matter and antimatter is broken, allowing the matter-making process to run faster than the antimatter-making process.

Finally, we have condition 3: the universe needs to have been out of thermal equilibrium when these processes were running. By definition, a system that's in thermal equilibrium isn't changing, usually because all processes are running forwards and backwards at the same rate. Therefore, we need to find a time in the universe's history when things were out of balance, allowing the matter-making process to run forwards faster than it ran backwards.

One of the great missions of theoretical and experimental physics over the past few decades has been to find a recipe that satisfies all three of Sakharov's conditions at the same time. There are several speculative ideas on the market, but we'll focus on what are generally regarded as the two most promising candidates. Although we don't know which one is right yet, physicists around the world are working hard to bring the pieces of the puzzle together so that one day, perhaps, we might learn the recipe for everything.

THE MIRROR WORLD

When you look in a mirror, in all likelihood the person you see is subtly different from the one your friends, colleagues, and lovers know and, presumably, love. Perhaps your nose bends slightly to the left, or your smile is a little lopsided, charmingly so, of course. You shouldn't feel bad about this. Even Hollywood actors and supermodels are never totally symmetrical. We are all at least a little bit wonky.

Unlike us imperfect humans, for a very long time the laws of physics were thought to be endowed with perfect symmetry. Reflect the universe in a mirror and it was assumed that its mirror image would be indistinguishable, with every process carrying on exactly as before. The idea that nature should have a left- or right-handed bias seemed nonsensical. This was such a basic assumption that no one ever really gave it much thought, that is until a stunning experiment carried out by the Chinese American physicist Chien-Shiung Wu.

Wu's famous experiment, which she performed at the US National Bureau of Standards in 1956, produced a truly earth-shattering result: the weak force seemed to be wonky. Or, more precisely, the weak force seemed to prefer left-handed particles over right-handed ones.

It may seem strange that a particle can be left- or right-handed, but it all comes down to the fact that particles behave as if they are rotating, as described by the quantum mechanical property of spin. If you hold out both hands with your thumbs pointing upwards then the curl of the fingers on your left hand defines a left-handed rotation, while the fingers of your right hand define a right-handed rotation. In a similar way, the direction that a particle is spinning compared to the direction it's travelling determines whether it is right- or left-handed.

RIGHT-HANDED ELECTRON   LEFT-HANDED ELECTRON

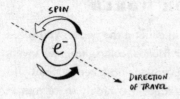 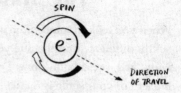

Wu discovered that electrons emitted by radioactive atoms of cobalt-60 tended to be left-handed more often than they were right-handed. This was a truly shocking result. When the famous quantum physicist Wolfgang Pauli heard about Wu's experiment he exclaimed, 'That's total nonsense!' However, it wasn't nonsense;[*] the weak force really does violate mirror symmetry, which physicists refer to as 'parity violation'.

The ultimate reason for parity violation is that the weak force interacts more strongly with left-handed particles than right-handed ones – in other words it 'prefers' left-handed particles. In fact, if the particles involved are massless, the weak force *only* interacts with left-handed particles. This isn't true for the electromagnetic or strong forces, both of which have no preference for left or right.

Faced with the loss of mirror symmetry, physicists suggested that perhaps order could be restored by adding another symmetry into the mix: *charge* symmetry. Take the universe and reflect it in a mirror *and* flip the sign of all charges, so that positive becomes negative, negative becomes positive, and this new mirror universe should now look the same as the original one. Crucially, flipping charges around transforms particles, like electrons and protons, into their antiparticles, so in such a universe, left-handed particles would become right-handed antiparticles. In other words, you get a mirror universe made of antimatter!

If this combined charge-parity (CP) symmetry is exact, then it implies that just as the weak force prefers left-handed particles over right-handed ones, it ought also to prefer right-handed antiparticles over left-handed ones. This is indeed true as later experiments showed. If Wu had been able to get her hands on some anti-cobalt-60 atoms[†] and watched the antielectrons come whizzing out, she would have found more righties than lefties.

CP symmetry promised to restore order to the particle world.

---

[*] Wu had performed her experiment with meticulous skill, although, rather disgracefully in my view, she did not receive the Nobel Prize for her discovery, which instead went to the two theorists who had first suggested that the weak force might violate mirror symmetry.

[†] No one has yet managed to make such a large anti-atom. The heaviest so far is anti-helium, which was detected in the collisions produced by RHIC at Brookhaven in 2011.

However, it also posed a problem. If CP symmetry is exact, then it would have been impossible to make more matter than antimatter during the big bang and we would not exist.

Fortunately for us all, in 1964, an experiment performed at Brookhaven shattered the restored charge-parity symmetry. A small team led by James Cronin and Val Fitch were using Brookhaven's most powerful particle accelerator to study beams of particles called 'neutral kaons'. These exotic beasts come in two types, one made of a strange quark and a down antiquark, and its antiparticle made of a down quark paired with a strange antiquark. When Cronin and Fitch analysed the results of their experiments, they were stunned to discover that these particles were decaying in a way that broke the sacred principle of CP symmetry.

If Wu's discovery of parity violation sent tremors through physics, the discovery of CP violation caused an earthquake. Cronin and Fitch's findings were so surprising that many theorists went to great lengths to explain them away, but before long, hard experimental evidence proved the case beyond doubt: CP symmetry is not an exact symmetry of nature. Reflect the universe in a mirror and flip the charges of all the particles, and the mirror anti-universe you find will look ever so slightly different from the one we live in. Nature's mirror is warped.

The discovery of CP violation makes it at least possible to imagine a recipe for making more matter than antimatter, but it's not enough on its own. For one, it is still far from clear whether there is enough CP violation in nature to account for the preponderance of matter that we see in the world around us. We'll come back to this thorny issue shortly. But more to the point, we've only satisfied one of the three Sakharov conditions. We still need to find a way to actually make more particles than antiparticles, something that we have *never* seen happening in the real world. However, such a process has been at least imagined, and remarkably, it doesn't require any new particles or forces beyond the ones we know about. Unfortunately, what it does require is some very, very hard maths . . .

ENTER THE SPHALERON

'I remember this flash of inspiration. I was walking home from the Maths Institute in Oxford, where I was visiting for three months in 1983, back to my apartment along Banbury Road, I remember almost exactly where it was and suddenly, I realized, Wow! I've got it!'

I met Nick Manton over a cup of tea at Cambridge's Department of Applied Mathematics and Theoretical Physics on a damp October morning. We chatted surrounded by a couple of dozen theoretical physicists enjoying their midmorning coffees, watched over by a bronze bust of their late colleague, Stephen Hawking. When Nick noticed me gazing enviously at the impressive spread of cakes and biscuits on offer (at the Cavendish we have to bring in our own), he told me that the largesse was thanks to Hawking, who had set aside a legacy to fund the group's eleven o'clock coffee break in perpetuity. That's one sure-fire way to make certain you're remembered fondly by your colleagues.

I was there to try to get my head around a strange set of objects that Nick, as a young researcher almost forty years earlier, had discovered hiding in the equations of the standard model. What had begun as a bit of a theoretical curiosity opened up one of the only viable ways we know of making more particles than antiparticles. These strange objects are called 'sphalerons' and they just might be the reason that everything in the universe exists.

'You know, I'd been sort of mulling over this thought, "What is this possible solution in the electroweak theory?" And it was based on someone else's work, someone else had found an unstable solution and I was thinking, "Is that idea relevant? How can it be applied to what I'm interested in?" And then suddenly in a flash of inspiration, I saw it.'

What Nick saw is almost impossible to describe to someone without advanced training in quantum field theory, but back in his office, he nonetheless patiently took me through the logic of the idea, covering a blackboard with intersecting spheres and circles, symbols representing the Higgs field, and towers of particle energy levels in a valiant attempt to convey the core of the idea. His explanation was so clear and methodical that while he was speaking, I really did think I was following, but as soon as I left his office, I could feel my tenuous understanding evaporating like the memory of a dream.

Sphalerons are features of the same electroweak theory that describes the origin of the electromagnetic and weak forces and the Higgs field. When Nick was walking through Oxford in 1983, the W and Z bosons had just been discovered at CERN, putting electroweak theory on solid experimental ground for the first time. He had been thinking deeply about the electroweak equations, and about a particular way to solve them that gave rise to an unstable arrangement of many fields, moving together *collectively*. Since this collective motion of several fields was highly unstable, Nick coined the term 'sphaleron', from the Greek *sphaleros*, meaning 'ready to fall'.

The first thing to say about a sphaleron is that it is not a particle. A particle, like an electron or a Higgs boson, is a single quantum field wobbling backwards and forwards around its mean value, like a single note played on a guitar string. A sphaleron, on the other hand, is something more subtle. It is still made of quantum fields, but instead of being made of one field, it is a baroque mixture of the W, Z, and Higgs fields all moving together as one, an orchestra of quantum fields playing in melodic unison.

This collective motion of the W, Z, and Higgs fields gives rise to an object that can do something remarkable – it can convert antiparticles into particles and vice versa. A sphaleron can act as a matter-making machine; all you need to do is feed some antimatter and a spray of ordinary matter particles will emerge.

This miraculous ability makes sphalerons the *only* objects within the framework of the standard model that can break the perfect balance between particles and antiparticles, giving them a uniquely important role in our understanding of the origin of matter. In all other cases, finding a recipe for matter means introducing new exotic particles, but what makes sphalerons so appealing is that you can do it with just the good old W, Z, and Higgs fields. 'You don't need any extra bells and whistles,' as Nick put it.

The question is, do such bizarre objects really exist in nature, and if so, what would they be like? Well, even though a sphaleron is not a particle, it would look a lot like one. It would have a well-defined position in space, a mass, and a size. What's more, you can use the equations of the standard model to calculate how big and heavy a sphaleron ought to be. The answers you get are staggering.

A sphaleron would be about 10^{-17} metres across – that's a hundredth of a thousandth of a trillionth of a metre – with a volume a million times smaller than a proton. On the other hand, this fantastically tiny object would have stupendous mass, weighing in at a whopping 9 TeV (trillion electron volts), which is almost ten thousand times heavier than a proton and far, far more massive than the heaviest particle we have ever detected.

Its minuscule size and gargantuan mass make a sphaleron 10 billion times denser than a proton. To put it another way, a teaspoon of sphalerons would weigh twice as much as the Moon. Such incredible densities are thought to be beyond what even the Large Hadron Collider can create in its most violent particle collisions. What's more, making a sphaleron isn't as simple as bashing particles together extremely hard; the energy has to go into the right set of fields in the right order. Trying to make a sphaleron at the LHC is a bit like firing a barrage of tennis balls at an orchestra and expecting to hear Beethoven's Ninth Symphony. You need W, Z, and Higgs fields all to play together as one, and the chances of that happening randomly in a particle collision are thought to be very slim indeed.

However, there was a time in the universe when such extreme and yet specific conditions existed – in the first trillionth of a second of the big bang. Back then, the plasma filling the universe was incredibly dense, dense enough to make sphalerons in large numbers. And what's more, the primordial plasma would have flowed collectively, like currents moving through the ocean, making the chances of getting the W, Z, and Higgs fields all moving together in the right way far more likely than at a particle collider.

The presence of sphalerons in the early universe provides an almost unique mechanism for producing more matter than anti-matter. Indeed, the most promising recipes on offer today for making matter all use sphalerons in one way or another. The trouble is, how do we really know if these things exist in nature?

Well, for one, sphalerons appear to be an inevitable consequence of electroweak theory, and given that the W, Z, and Higgs were all discovered just as predicted, theorists are fairly confident that sphalerons must exist too. What's more, despite the initial gloominess over the prospects of detecting them at colliders, some recent theoretical work has shown we may yet have a chance.

Calculations suggest that random fluctuations in the collisions at the LHC might just allow a sphaleron to form once in a blue moon. They would then immediately decay into a spray of ten different matter particles, a fairly unique signature that physicists at ATLAS and CMS would have a decent chance of spotting among all the other subnuclear detritus.

An even better prospect might come from collisions between heavy nuclei, like the gold–gold collisions used to study quark–gluon plasma at Brookhaven. These mighty smash-ups involving hundreds of protons and neutrons generate absolutely stupendous magnetic fields over very short distances, whose powerful pull might be enough to get the W, Z, and Higgs fields all wobbling together in just the right way to make a sphaleron. No one has seen one yet, but maybe, just maybe, the standard model's most bizarre objects may yet reveal themselves to us. If and when they do, we will have another of the three crucial ingredients for making matter in the early universe.

A RECIPE FOR QUARKS

So far, things are going pretty well. The standard model appears to satisfy two of the three Sakharov conditions needed to make more matter than antimatter during the big bang. Sphalerons provide us with a way to convert between particles and antiparticles, while we know from experiments that the weak force breaks the symmetry between quarks and antiquarks. All we need now is a time in the universe's history when it was out of equilibrium and we could be in business.

Amazingly, the discovery of the Higgs boson implies that just such an event occurred around a trillionth of a second after the big bang. This was the moment when the Higgs field switched on, giving mass to fundamental particles and changing the basic ingredients of the universe beyond recognition. This crucial event could well be the reason that everything in the universe exists.

Before the Higgs field switched on, the fundamental particles of nature looked very different from how they do today. The quarks and electrons that would go on to make up ordinary matter had no mass and zipped around at the speed of light, interacting with each other through a single unified electroweak force. However, after around a trillionth of a second the rapidly expanding universe's temperature dropped below a critical threshold (around 100 GeV), causing the Higgs field to rise to a constant value throughout the entire universe, giving mass to quarks and electrons and causing the electroweak force to break apart into separate electromagnetic and weak forces.

This event is known as the 'electroweak phase transition', and crucially, for our story, it satisfies the third and final Sakharov condition – a time when the universe was out of equilibrium. Combined with the experimental discovery that charge-parity symmetry (the symmetry between matter and antimatter) is broken by the weak force and the theoretical prediction of the existence of sphalerons that can convert antimatter into matter (and vice versa), the electroweak phase transition could have been the moment when nature's scales were tipped in ordinary matter's favour, ultimately giving rise to the material universe.

How could this have happened? Well, as the term 'phase transition' suggests, this was a moment when the universe underwent a rapid change of state, similar to more familiar phase transitions like steam cooling to form liquid water or water freezing into ice. Making matter during the electroweak phase transition depends on exactly how the phase transition happened – specifically, whether it was smooth and even or sudden and uneven.

A smooth and even transition is no use for making matter, as sphalerons would have converted particles into antiparticles and antiparticles into particles at the same rate, preserving the perfect balance between matter and antimatter. However, if the electroweak phase transition happened unevenly, then making more matter than antimatter becomes possible.

There's no getting away from the fact that this process is a little complicated, but we are talking about the recipe for everything here. I'll take it step by step.

As the universe cools, the Higgs field turns on in some places before others, causing bubbles to form in the searing-hot plasma that fills the universe. Inside these bubbles where the Higgs field is on, quarks and electrons have gained mass and the weak and electromagnetic forces have come into existence. Outside these bubbles, the Higgs field is still off, particles have no mass, and there is still a single unified electroweak force.

You can picture these bubbles as like droplets of liquid water condensing out of a cloud of steam; just as light reflects off the surface of a water droplet, quarks and antiquarks reflect off these bubbles. Some of the quarks and antiquarks zooming about in the external plasma will collide with a bubble, either passing inside or bouncing off back into the surrounding plasma.

Thanks to the fact that the weak force breaks charge-parity symmetry – that is, it interacts with particles and antiparticles slightly differently – the chances of an antiquark bouncing off the bubble walls are slightly higher than for a quark, while a quark has a better chance of passing into the bubble. As a result, more antiquarks end up outside the bubbles while there are more quarks inside. There are still equal numbers of quarks and antiquarks overall, but they're now unevenly distributed.

This is where sphalerons play their starring role in the story. Sphalerons can't exist inside the bubbles where the Higgs field has switched on, but outside the bubbles where the Higgs field is off they are being produced all the time. The fact that there are sphalerons outside the bubbles but not inside is crucial. Outside the bubbles, sphalerons gobble up the extra antiquarks and convert them into quarks, while the excess quarks inside the bubbles are safely out of the sphalerons' reach and stay as they are. The result is that, for the first time in the (admittedly still extremely brief) history of the universe, there are now more quarks than antiquarks.

While all this is going on, the bubbles are getting bigger and bigger, swallowing the newly made quarks and saving them from being turned back into antiquarks by the sphalerons. A tiny instant after the phase transition began, these bubbles start to collide with one another, merging into bigger and bigger regions until eventually the entire universe has been filled with the new 'Higgs on' state. This kills off the sphalerons, preventing any more conversions and freezing the imbalance between quarks and antiquarks forever. This tiny imbalance is enough to give matter the edge during the great annihilation, around a microsecond later, leaving just enough to form everything we see in the world around us.

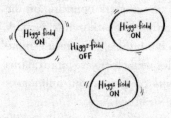

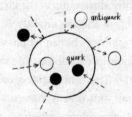

1. The Higgs field starts turning on, forming bubbles.

2. Charge-parity violation means more antiquarks bounce off the bubbles than quarks, creating an excess of antiquarks outside the bubbles.

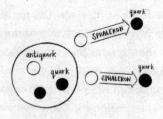

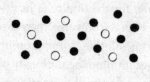

3. Sphalerons outside the bubbles convert antiquarks into quarks.

4. Bubbles expand and merge, leaving more quarks than anti-quarks.

This rather miraculous process is known as 'electroweak baryo-genesis',* which is just a fancier and more compact way of saying 'making quarks when the Higgs field turned on'. One thing that makes it extremely appealing is that it is experimentally testable. That may seem like a low bar for a scientific theory, but as we'll see, many of the ideas that we encounter as we move closer and closer to the big bang are more or less impossible to test directly due to the vast energies involved. The electroweak phase transition,

---

* *Baryo-* refers to baryons, the class of particles made from three quarks that includes protons and neutrons, and -*genesis* meaning 'origin'.

on the other hand, took place when the temperature of the universe was around 100 GeV, well within the collision energy of the LHC, which can smash protons together at 14,000 GeV. That means that the LHC should be able to recreate the particles and phenomena that were involved and test whether it really was how matter got made in the early universe.

However, the idea immediately runs into some problems if we only use the ingredients provided by the standard model. A major stumbling block is that the amount of matter–antimatter asymmetry that we've measured so far seems to be way too small to get the process to work. What that effectively means is that the probabilities for quarks and antiquarks to bounce off those Higgsy bubbles would be too similar, and you wouldn't get a big enough excess of quarks building up inside the bubbles.

Another serious issue has to do with the electroweak phase transition itself. Now that we know the mass of the Higgs boson, theorists can plug it into their models and calculate how the phase transition would have happened. What you find is that instead of taking place unevenly in bubbles, the phase transition would have happened evenly and smoothly throughout all of space, and without bubbles to separate quarks from antiquarks the whole process becomes impossible.

Still, all is not lost. Both of these problems can be fixed as long as there are new quantum fields beyond the ones we've found so far. These quantum fields would have to break the CP symmetry between quarks and antiquarks and also change the way the Higgs field behaves to allow bubbles to form as it switched on during the big bang. Encouragingly, these fields should be detectable in experiments.

The obvious place to look for new quantum fields is the Large Hadron Collider. If they exist, then the LHC should be able to hit them hard enough to make them wobble and create some of the particles that go with them, which could then be spotted by the gigantic ATLAS and CMS experiments. Meanwhile, at LHCb, we've spent the last decade searching for new signs of matter–antimatter asymmetry by studying various types of exotic quarks. The 'b' in LHCb stands for 'beauty', the name of a heavy cousin of the more familiar down quark that's found inside protons and neutrons. One of the main goals of LHCb is to study the billions upon billions of beauty quarks produced by the LHC and to see if we can find differences between how the beauty quark and antiquark decay.

Unfortunately, so far ATLAS and CMS haven't seen any signs of new particles being created in their collisions, although they may still turn up as the LHC ramps up its collision rate over the next decade. At LHCb we've seen plenty of evidence for beauty quarks breaking matter–antimatter symmetry, and more recently we've even caught charm quarks (heavy cousins of the up quark) getting in on the symmetry-breaking act. But alas, the amount of matter–antimatter asymmetry is still way below the level needed to explain the dominance of matter in the universe.

If results coming out of the LHC aren't particularly encouraging for electroweak baryogenesis fans, then things look even gloomier when you consider the results of a totally different, and far less expensive, set of experiments. Rather wonderfully, the strongest arguments against electroweak baryogenesis come from measurements of the shape of the electron, including the decidedly low-frills experiment in the basement of Imperial College that we encountered a while back.

If new symmetry-breaking quantum fields exist, then they should gather around the electron and squash it from a perfect sphere into something more like a cigar. The fact that the most sensitive electron-shape-measuring experiments in the world have all found the electron to be as round as round can be is starting to put serious pressure on the existence of these new quantum fields.

So things aren't looking all that rosy for this particular recipe for making matter. Of course, there is still room for new quantum fields to show up at the LHC or in the ever-more-precise measurements of the electron's shape being planned for the near future. So while it's not game over quite yet, theorists are increasingly looking elsewhere to explain the existence of matter in the universe. The most popular alternative involves the most elusive particles that we've discovered so far: neutrinos.

MATTER MADE BY GHOSTS

Deep under Mount Ikeno in central Japan is one of the most spectacular human-made spaces in the world. Housed a kilometre belowground in an old zinc mine is a soaring cylindrical tank containing 50,000 tonnes of ultrapure water, big enough to swallow the Statue of Liberty whole. Its walls, floor, and roof are covered with thousands upon thousands of gleaming golden orbs, electrical eyes that watch for faint flickers of light emerging from the dark water, the telltale sign of an arriving neutrino. This is Super-Kamiokande (Super-K), the world's largest neutrino observatory, and it may have given us a crucial clue to the origin of matter.

In April 2020, the 150-strong international team that operates Super-K reported the first hint that neutrinos can also break the charge-parity symmetry that relates matter to antimatter. If confirmed by more precise measurements, this is a huge deal. So far, only quarks have been caught breaking charge-parity symmetry through their interaction with the weak force. If neutrinos can do it too, then it opens up a second potential way to make matter in the first instant after the big bang.

To understand why Super-K's result is so significant, we first need to quickly recap what we know about neutrinos. Neutrinos are the most abundant matter particles in the universe, and yet we are almost completely unaware of them thanks to the fact that they have no electric charge and only interact with ordinary atoms through the weak force. As a result, they can pass through solid objects, including planets and stars (not to mention Italian mountain ranges), as if they weren't there, which sends science writers scouring their thesauruses in search of new spooky adjectives once they've used 'ghostly' and 'elusive' more than a couple of times.*

These, ummm, wraithlike neutrinos come in three types or 'flavours', the electron, muon, and tau, each of which is paired up with an electrically charged particle. Fire a beam of electron neutrinos at some atoms with sufficient energy and a few of them will convert into electrons. Do the same thing with muon or tau neutrinos and you'll get negatively charged particles called, unsurprisingly, 'muons' and 'taus', which are heavy, unstable cousins of the electron. Together, the three neutrinos and the electron, muon, and tau particles form a family of six fundamental particles called 'leptons'.

THE LEPTONS

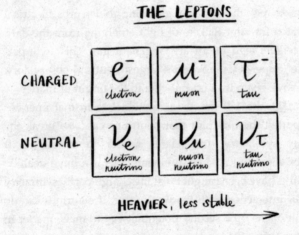

Another thing that we used to believe about neutrinos is that they were completely massless. That is until a major discovery made by Super-K more than twenty years ago. In 1998, scientists there announced they had found evidence of muon neutrinos morphing into tau neutrinos as they travelled through the Earth. This phenomenon is known as 'neutrino oscillation', and all three

---

* Guilty as charged.

neutrinos can do it. Produce a beam of pure electron, muon, or tau neutrinos and plonk a detector a few kilometres away and you'll discover that a fraction of them have shape-shifted into the other two flavours en route.

While interesting in its own right, the real significance of Super-K's discovery was that neutrinos could only pull off this quantum mechanical Jekyll and Hyde act if they had mass. Until that point, no experiment had seen any direct evidence that neutrinos had anything but fantastically tiny masses, so it had not unreasonably been assumed that they had no mass at all. In fact, neutrinos do have masses, they're just so stupendously minuscule that we haven't been able to measure them. All we can say is that they must have masses smaller than 0.5 electron volts, more than a million times lighter than an electron. The question is, why are their masses millions of times smaller than the other matter particles?

The most popular answer to this question is known as the 'se mechanism', which, as the name kind of implies, counter the lightness of the ordinary neutrinos by proposing t of three additional extremely heavy neutrinos. You can heavyweight neutrinos as being like rugby prop fo at one end of a theoretical seesaw, while the ordinary are like ballerinas stranded high in the sky at the

Now just in case you're scratching your head a are correct in thinking that I haven't actually neutrinos have such tiny masses apart from so analogies to do with seesaws. Unfortunately, tho so involves too much hard maths to go into here. n they point is that if these heavyweight neutrinos exi the big lutely clear, so far we have zero evidence that could have been responsible for making m utrinos, these bang.

To explain the incredible lightness of ord rs, with masses heavy neutrinos would have to be absolute of between a billion and a thousand trillion protons (that's between 10^9 GeV and 10^{15} GeV), far, far heavier than any particle we have ever seen to date and at least one hundred thousand times higher in energy than the LHC can reach. However, although it's impossible to create them in colliders today due to their enormous masses, they could have been made in the furious conditions of the very, very early universe when, as we've seen, temperatures were unimaginably high.

It is these heavy neutrinos that may have been responsible for the imbalance of matter over antimatter in the universe. As the universe expanded and cooled down, there wouldn't have been enough energy left to make more of them and they would have started to decay into Higgs bosons and ordinary leptons (that's the three light neutrinos and the electron, the muon, and the tau).

If these heavy neutrinos break charge-parity symmetry, then it is possible for them to decay into antileptons more often than leptons, leading to a universe with more antiparticles than particles. Now, this may not sound particularly helpful, as surely we want a universe with more particles than antiparticles, and not the reverse? This is where our old friend the sphaleron rides in to save the day.

Remember we said that sphalerons can convert antiparticles into particles? Well, a bit later in the universe's history (although we're talking within the first trillionth of a second) sphalerons would converted all these excess antileptons into ordinary matter, including quarks and electrons, giving rise to the basic ts that would go on to form everything we see around us

forgiven for thinking that this recipe for matter is speculaera speculation, and you'd be right. If these heavy This then they're way out of reach of any particle accelthis recipe agine today. So how can we possibly test this idea? uper-K comes in. One of the key ingredients of hese heavy neutrinos broke matter–antimatter symmetry when they decayed just after the big bang. Given that we can't get our hands on these heavy neutrinos, there's no way to test that directly, but if we can catch ordinary neutrinos breaking matter–antimatter symmetry then that would be a vital clue that their heavy cousins can do it too.

The Tokai to Kamioka (T2K) experiment begins 295 kilometres to the east of the vast Super-K neutrino observatory, at Tokai on the Pacific coast. Here, a powerful particle accelerator slams protons into a graphite target, creating a shower of particles. Some of these particles decay into neutrinos, which tunnel directly through the Earth towards the Super-K observatory on the other side of the country. Thanks to their ghostly properties, the neutrinos are completely unbothered by 295 kilometres of rock, with only a small fraction being absorbed as they travel.

Critically, T2K has the ability to produce beams of either muon neutrinos or their antimatter versions, muon antineutrinos. As they travel through the Earth towards Super-K, the neutrinos start to morph into other flavours, and by the time they arrive at the observatory a certain fraction of them have converted into electron neutrinos. A tiny fraction of these electron neutrinos smack into a water molecule in Super-K's huge tank, producing electrons that emit a flash of light as they whizz through the liquid. By counting the number of electrons created, T2K can measure the probability that a muon neutrino morphs into an electron neutrino. By switching to a beam of muon antineutrinos, it can also measure how often a muon antineutrino turns into an electron antineutrino.

If neutrinos respect matter–antimatter symmetry, then T2K should measure equal probabilities for muon neutrinos to turn into electron neutrinos as for muon antineutrinos to turn into electron antineutrinos. But in April 2020 the team announced that they had seen compelling evidence that neutrinos were more likely to switch flavour than their antimatter versions. What's more, the numbers they saw suggest that neutrinos don't just break matter–antimatter symmetry a little bit, they break it by the maximum amount possible.

This result is really quite exciting. If neutrinos actually do break matter–antimatter symmetry it suggests that their heavy cousins could have done the same thing at the start of the universe, sowing the seeds for creation of ordinary matter. That said, the results aren't yet precise enough to be absolutely certain of the effect. Future upgrades to T2K and giant new neutrino experiments in Japan and the United States should be able to clear up the picture in the coming years.

However, even if T2K's results are confirmed, we'll still only have suggestive evidence that heavy neutrinos were responsible for making matter during the big bang. The heavy neutrinos themselves will most likely be forever beyond our reach. Here we run into a problem that will become increasingly frustrating as we push closer and closer to the big bang. The energies that were present at the dawn of time were far, far higher than we could achieve even in particle physicists' most feverish dreams, meaning that these theories are ultimately only loosely tethered to the firm ground of experimental observations. This is one of the main attractions of the previous recipe for making matter when the Higgs field switched on – it can actually be tested in experiments either today or in the near future. On the other hand, if heavy neutrinos were responsible for making matter during the big bang, we may never know for certain.

But we shouldn't be too gloomy. If the history of science teaches us anything it is that many of the biggest breakthroughs begin with an unexpected experimental result that radically challenges accepted principles or assumptions. Hardly anyone expected nature to break mirror symmetry before Wu's experiment showed that it did. The fact that quarks can break matter–antimatter symmetry too was a bolt from the blue. An experiment with the potential to throw up just such a result is currently going on at CERN. It is the stuff of science fiction, a place where a small team of scientists make, store, and study the most volatile substance in the universe.

THE ANTIMATTER
FACTORY

On a hot late-summer morning I found myself waiting outside a large but otherwise nondescript metal warehouse deep in the sprawling CERN site. I have always assumed that CERN's buildings were numbered by someone with a sense of humour, given that they seem to be scattered more or less at random across the 200-hectare laboratory, which can make finding an unfamiliar building an interesting challenge. Luckily, locating Building 393 turned out to be a fair bit easier than I'd feared thanks to a giant blue sign bolted to one of its corrugated walls, proclaiming 'ANTIMATTER FACTORY'.

As a result, I was fifteen minutes early to meet Jeffrey Hangst, spokesperson of the ALPHA experiment, and so did my best to look innocent as I loitered by the security door. After all, if Hollywood has taught us anything about particle physics it's that unscrupulous individuals will do almost anything to get their hands on some antimatter.

Bang on time, Jeffrey came striding down the road towards me. Tall, lithe, and sporting a black T-shirt, shades, and short grey stubble, he looked more rocker than physicist; the only giveaway was the CERN lanyard hanging around his neck. As I was about to find out, ALPHA is a pretty rock 'n' roll experiment. Inside the Antimatter Factory we were hit by a wall of noise: humming machinery, the rhythmic chirrup of compressors, and the occasional blast of a siren as a bridge crane slid across the roof high above our heads. Growing up in a small town in Pennsylvania's steel country, Jeffrey had been told that if he didn't study hard and get into university he'd end up at the local steelworks, so it's a strange quirk of fate that he goes to work in a factory each day, albeit one of a rather different kind.

Here at the Antimatter Factory, Jeffrey and his fifty collaborators on ALPHA manufacture and study atoms of antihydrogen, the simplest atom of antimatter. This is no mean feat. Since there aren't any handy reserves of antimatter in our cosmic neighbourhood, ALPHA has to create its anti-atoms from scratch by carefully mixing positively charged antielectrons with negatively charged antiprotons. And even once you've got some, holding onto antihydrogen long enough to study it is a seriously difficult task. After all, how do you contain a substance that annihilates immediately on contact with ordinary matter? When I carelessly referred to ALPHA as a 'detector' Jeffrey flashed me a scornful look, 'ALPHA is *not* a detector. Detection is just a tool for us. The real art is learning how to trap neutral anti-atoms. That's really hard. We make antihydrogen so that when it's born, it's trapped and we're the only experiment in the world that knows how to do this. So, when you call it a detector, it really gets my hackles up.'

Trapping antielectrons or antiprotons is relatively straightforward. Since they're electrically charged, a judicious arrangement of electric and magnetic fields can keep them floating safely in the centre of a vacuum vessel, but once they combine to make neutral antihydrogen, you're in a whole different ball game. With no net electric charge, antihydrogen atoms are far, far harder to manipulate. ALPHA was the first experiment in the world to crack the problem. In late 2010, they managed to hang on to thirty-eight antihydrogen atoms for around a sixth of a second; today they can store a thousand more or less indefinitely.

As Jeffrey himself put it, when he started out at CERN a couple of decades ago this would have seemed like science fiction. In fact, it literally was. In 2008 he showed Ron Howard and Tom Hanks around the experiment while they were shooting the movie adaptation of Dan Brown's thriller *Angels and Demons*, whose plot centres around a nefarious organization stealing a canister of antimatter from CERN in a dastardly plot to blow up the Vatican. In reality, if you could gather together all the antihydrogen that ALPHA has ever trapped, it wouldn't be enough to blow up a housefly, let alone a city.*

ALPHA isn't in the antimatter bomb business. Its real goal is to make exquisitely precise measurements of the properties of the spectrum of antihydrogen atoms. Just like in ordinary hydrogen, the antielectron orbits the antiproton in fixed quantum energy levels, jumping from one orbit to another by absorbing or emitting photons. By measuring the frequencies of these photons and comparing them against the spectrum of ordinary hydrogen, Jeffrey and his fifty collaborators test for breaks in the symmetry between matter and antimatter, which maybe, just maybe, could provide a clue to how matter came to exist in the universe.

The symmetry that relates an ordinary hydrogen atom to its antimatter version is known as 'charge-parity-time (CPT) symmetry.' Charge-parity symmetry we've already met; it involves flipping particles into antiparticles (and vice versa) and then reflecting the universe in a mirror, so that left becomes right and right becomes left. We know that nature breaks CP symmetry in the way quarks decay and possibly also in neutrino oscillations. However, if you add in time-reversal (T) symmetry, where in effect you reverse the directions that particles are travelling in, then it is believed that the laws of nature will ultimately remain unchanged. Or to put it another way, an antimatter universe, reflected in a mirror, where time runs backwards, would look completely identical to the one we live in.

CPT symmetry is so fundamental to quantum field theory that most theorists think that it must be unbreakable and that Jeffrey and his colleagues are on a hiding to nothing. However, very few people expected either parity or charge-parity symmetry to be

---

* In 1999, NASA estimated that a single gram of antihydrogen (which is about what you'd need for a city-destroying bomb) would take the age of the universe to make and set you back around $62.5 trillion, so it would be more efficient for the Illuminati to buy the Vatican outright and then hire an army of builders to take it apart brick by brick.

violated before experiments showed that they are. If CPT symmetry turned out to be broken too, it would be totally revolutionary and shake quantum field theory to its foundations. Or as Jeffrey put it, 'The theory guys are like, "Okay it's CPT, fuck it, CPT's good," but these symmetries are always good until they're not. The argument for CPT as an immutable thing assumes that quantum field theory is the final word, and that's incredibly arrogant. There are so many things we don't know. I just refuse to accept when people say we know CPT is a law because again and again they've been wrong. I think as an experimentalist you have to filter that out and just do the best experiment you can do.'

But even putting the potential theoretical implications aside, Jeffrey is clearly in this for the love of the challenge itself. 'For me it's how could you *not* do this?' he said. 'It's become possible in my lifetime to make and hold on to antimatter atoms. That's incredible! If you could have seen how this started, it was a kind of ragtag bunch of people and nobody believed we would ever make antihydrogen, nobody believed we would ever capture it if we could make it, and nobody believed if we could do all that we would ever have enough. Now I can measure a single spectroscopic line in antihydrogen in one day. It's routine now for us.'

The unquestionable coolness of what the team at ALPHA are doing has attracted a fair amount of celebrity attention. On our way down to the factory floor to see the experiment, Jeffrey pointed proudly to a whiteboard covered with signatures of the big names who have come to see ALPHA. Alongside a signed photo of Ron Howard were the signatures of Roger Waters, David Crosby and Graham Nash, Jack White, and the members of Muse, Slayer, Metallica, the Pixies, and the Red Hot Chili Peppers. There was clearly a strong rock bias. Jeffrey himself plays guitar in a band that performs annually at CERN's Hardronic music festival (yes, that really is a thing and it's *quite* the event) and he is pretty selective about whose name goes up. 'Anyone you'd like to get?' I asked. He replied without hesitation: 'David Gilmour from Pink Floyd.'

Down another staircase and we were standing next to the experiment itself, an anarchic-looking arrangement of cables, pipes, electronic readouts, flashing lights, metal framework, and glistening insulating foil. At its centre was a gleaming stainless-steel vessel – the antihydrogen trap itself.

To make antihydrogen, you first need antiprotons, which are created when one of CERN's large particle accelerators fires protons into a target, creating a shower of particles and antiparticles. The antiprotons that come flying out are moving way too fast and chaotically to be used to make antihydrogen and so are first corralled and slowed by a unique machine called the Antiproton Decelerator, which then feeds them into ALPHA. However, even then the antiprotons still have far too much energy and have to be 'cooled' further by shooting them through sheets of aluminium foil and then mixing them with electrons. After all this the antiproton temperature has dropped from billions of degrees down to just 100 Kelvin (−173 degrees Celsius). Meanwhile, antielectrons produced by a radioactive source on the other side of the set-up are spiralled through a magnetic field to cool them down before being brought into the antihydrogen trap.

At first, the antiprotons and antielectrons are kept apart by an electric field, which is slowly adjusted to bring the two clouds of oppositely charged particles together. As they mix, neutral antihydrogen atoms are formed, which thanks to their slight magnetism, can be kept away from the walls of the trap by an extremely powerful magnetic field. Repeating this process over an eight-hour shift, the ALPHA team can now store up to one thousand antihydrogen atoms at a time, an incredible achievement that has taken two decades of painstaking design, innovation, and hard work.

Once the team have trapped a cloud of antihydrogen atoms, the final step is to measure their spectrum. Laser light is fired into the trap, and if the frequency is right, it will kick some of the antielectrons from the lowest energy levels into a higher one. If an antielectron gets promoted, another photon can then knock it out of the atom altogether, which causes the resulting antiproton to drift into the trap walls and annihilate. The particles released by the annihilation can then be detected, and by counting the number of annihilations the team can tell whether their laser is set at the right frequency to cause an antielectron to make a quantum jump.

After they first successfully trapped antihydrogen in 2010, ALPHA was completely rebuilt, and in 2016 the team finally managed to see a quantum jump in antihydrogen for the first time. Now they can do the same measurement in a single day and have determined the energy of the jump to a few parts in a trillion. 'I still don't believe that it's possible today,' Jeffrey told me with undisguised pride. 'We've surprised ourselves a lot.' So far, the spectrum of antihydrogen is in perfect agreement with ordinary hydrogen, and amazingly the precision of ALPHA's measurements are rapidly closing in on the measurements of ordinary hydrogen. 'We'll get close to hydrogen soon. Hydrogen has been measured to parts in ten to the fifteenth [a thousand trillion]. We're at parts in ten to the twelfth [a trillion] now, in only two years. They've had two hundred years!'

But ALPHA has another, potentially even more exciting measurement in the offing. Just to the side of the original experiment, Jeffrey pointed me to a tall metal structure rising several metres up from the factory floor. Inside was another version of ALPHA, this time mounted on its side so that it pointed vertically towards the roof. This, Jeffrey told me, is ALPHA-g, and its mission is to check whether antimatter falls up.

Considering that antimatter was discovered almost a century ago, it's amazing that we still don't know whether it's repelled by the gravity of ordinary matter. Again, most theorists think this is vanishingly unlikely, but until someone actually measures it, we can't be certain. The idea of ALPHA-g (where 'g' stands for 'gravity') is to produce antihydrogen and then release it and see which way it falls.

If antimatter did turn out to be repelled by ordinary matter it would have profound implications for our understanding of the universe. 'There's a lunatic fringe out there who claim that repulsive gravitation would solve everything: antimatter asymmetry, dark energy, dark matter. If that actually happens you could explain everything that we're missing,' Jeffrey told me as we stood staring up at the tank. Indeed, if antimatter is repelled by ordinary matter it would explain why we don't see any antimatter in the universe around us, as it would all have been forced to distant parts of the cosmos, removing the need to find a recipe for making more matter than antimatter in the big bang. 'There are a lot of guys that have an article under their mattress waiting for this measurement.'

In 2018, Jeffrey and the team were in a desperate race against time to get ALPHA-g ready to collect data before the CERN accelerator complex went into a two-year shutdown at the end of the year to allow for upgrade work on the LHC and its experiments. The ALPHA team knew that if they could just get the experiment switched on they would find out almost immediately whether antimatter falls up or down. 'I worked the hardest I've ever worked, from May to November, seven days a week, twelve-, fifteen-hour days, trying to make that measurement before the shutdown. We really wanted to nail that measurement. If I'd had an extra month, I'd have made it.'

In the end, they got agonizingly close but didn't quite get over the line, and so now they're working hard to get both ALPHAs into tip-top working order for the restart. Jeffrey and his colleagues are forever looking for ways to improve their experiment; 'If it works, fix it' is his motto.

Besides, the ALPHA team don't have the field to themselves. They share the Antimatter Factory with several other experiments, including two that are racing to measure the effect of gravity on antimatter. But Jeffrey is in no doubt about who is going to get there first: 'I'm really confident that we'll win.' After all, they were the first to trap antihydrogen and then the first to measure a quantum jump. You certainly wouldn't want to bet against them.

As an experimental physicist, I find ALPHA's approach to science really inspiring. Not only are the measurements they are making extraordinarily difficult, they are also testing principles that most respectable theorists would tell you are bound to be true. Jeffrey clearly takes the view that until a principle is put to the test, you can never be sure that it's right. As Richard Feynman, doyen of quantum field theory, famously said, 'It doesn't make a difference how beautiful your [theory] is, it doesn't matter how smart you are . . . If it disagrees with experiment, it's wrong.'

ALPHA is experimental physics at its purest: grabbing hold of the physical world and studying its most basic principles in the lab. It's the endless pursuit of precision, the joy of solving difficult problems, and the determination to be first. Jeffrey is clearly a man in love with his work. As he showed me blinking back into the bright summer sunshine, he told me, 'There's no other place like this. I have the only job in the world that I'm qualified for; it's either this or I'm playing guitar on the street. I got no choice. I remember that every day, I better not screw up.'

CHAPTER 12

The Missing Ingredients

'With these record-shattering collision energies, the LHC experiments are propelled into a vast region to explore, and the hunt begins for dark matter, new forces, new dimensions, and the Higgs boson.'

It was with these words that Fabiola Gianotti, spokesperson of the ATLAS experiment, had fired the starting gun for the search for new particles at the Large Hadron Collider, as the first high-energy protons smashed into one another on 30 March 2010. That day, the ten-thousand-strong CERN community was buzzing with optimism and anticipation, while theorists around the world waited impatiently for answers to questions that many had spent their careers puzzling over. After a very, very long wait, the greatest scientific instrument ever built had fired up, a once-in-a-generation opportunity to explore a new subatomic landscape where all kinds of strange and exotic objects were surely waiting to be discovered.

As Gianotti suggested, the Higgs boson was just one of them. In fact, for many particle physicists, perhaps even the majority, it was one of the new machine's less exciting quarries. The Higgs belonged to the old, established story of particle physics, the last missing piece of a standard model that had remained more or less unchanged since the late 1970s. Nima Arkani-Hamed, one of the world's leading particle theorists, was so confident that the Higgs would be found at the LHC that he had offered a year's salary to anyone willing to bet against it. Even many experimentalists saw finding the Higgs as a tick-the-box exercise, a bit of unfinished business from the twentieth century before the real journey into terra incognita could begin.

Despite all its successes – explaining the structure of matter, quantum fields, the forces of nature, and the origin of mass – we know that the standard model is at best incomplete, an echo of a deeper, more fundamental theory that we have yet to glimpse. For starters, many physicists regard the standard model as ad hoc, ungainly, even ugly. Take the forces. There are three in the standard model – the electromagnetic, weak, and strong – but why those three? We don't know. The electromagnetic and weak forces are unified, but the strong force is left hanging out on its own. Do all the forces unify at high energy? Again, we don't know. Perhaps most significantly of all, gravity is left out completely.

Things get even worse when you look at the matter particles. We are made of electrons, up quarks, and down quarks, which along with the electron neutrino form a quartet known as the 'first generation' of matter. We do not know why these four particles exist, we just observe that they do and put them into the theory by hand, like a botanist collecting flowers in a field. Why couldn't there have just been one? Or five? Or a hundred? On top of this, nature decided to include heavier, unstable copies of these four particles. There's a second generation of matter, which includes the muon, the muon neutrino, the charm quark, and the strange quark, and then a third set of even heavier and more short-lived particles: the tau, the tau neutrino, the top quark, and the bottom quark. Why three generations and not four, or a thousand? Again, we don't know.

THE STANDARD MODEL

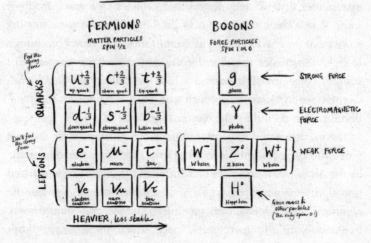

FERMIONS
MATTER PARTICLES
SPIN 1/2

BOSONS
FORCE PARTICLES
SPIN 1 OR 0

Feel the strong force

QUARKS

$u^{+\frac{2}{3}}$ up quark $c^{+\frac{2}{3}}$ charm quark $t^{+\frac{2}{3}}$ top quark

$d^{-\frac{1}{3}}$ down quark $s^{-\frac{1}{3}}$ strange quark $b^{-\frac{1}{3}}$ bottom quark

g gluon ← STRONG FORCE

γ photon ← ELECTROMAGNETIC FORCE

Don't feel the strong force

LEPTONS

e^- electron μ^- muon τ^- tau

ν_e electron neutrino ν_μ muon neutrino ν_τ tau neutrino

W^- W boson Z^0 z boson W^+ W boson   } WEAK FORCE

H^0 Higgs boson ← Gives mass to other particles (The only spin 0!)

HEAVIER, less stable →

Since the standard model was first put together, many have yearned for a deeper, more elegant theory, where all its apparent arbitrariness would be explained by some single, unifying principle. As we saw a couple of chapters back, the forces appear to arise because of symmetries in the laws of nature. Perhaps the standard model is only a corner of a larger, more symmetrical structure, like fragments of glass fallen from a great stained-glass window in some medieval cathedral. Only by finding other missing pieces will the full beauty and majesty of nature's fundamental laws be revealed. Of course, nature doesn't have to abide by our sense of what is beautiful. The desire for a more unified theory is really just an aesthetic one, even if unification and simplicity have been powerful guides in the past. But aesthetics aside, there are solid, observational reasons for believing that we are missing something big.

We've already seen that new quantum fields are needed to explain how matter got made in the fearsome heat of the big bang, but some of the greatest challenges to the standard model come not from particle physics, but from astronomy. During the twentieth century, observations of the heavens began to hint that there is far more to our universe than meets the eye. In the 1930s, the Swiss astronomer Fritz Zwicky found that galaxies in a giant group of more than a thousand known as the Coma Cluster were moving so fast that the gravity of the visible stuff shouldn't be strong enough to hold the cluster together. He suggested that the cluster must contain some invisible matter, which he called '*dunkle Materie*', German for 'dark matter', which was generating an extra gravitational pull and binding the cluster together.

The existence of dark matter remained controversial for the best part of forty years until a series of exquisite measurements made by the American astronomer Vera Rubin in the 1970s. Rubin showed that stars orbiting in spiral galaxies, including our nearest galactic neighbour, Andromeda, were moving so fast that they should break free and fly off into intergalactic space. Again, there simply didn't appear to be enough matter in the galaxies to generate the gravity needed to hold the stars in orbit.

Although Rubin's observations were received with scepticism at first, over the coming years it became clear that the effect was real. One possibility, still explored by a fringe group of physicists today, was that Newton and Einstein had got their theories of gravity wrong, and that gravity was stronger at long distances than first supposed. However, by far the more popular explanation is that almost every galaxy, including our own Milky Way, sits at the centre of a vast cloud of invisible dark matter, whose gravitational pull keeps the stars on their orbits. This unseen material is known as dark matter because it doesn't emit, absorb, or reflect light, making it completely invisible to our telescopes. However, astronomers can infer its presence by the way its gravity pulls on stars, galaxies, and light as they move through the cosmos, a bit like a poltergeist moving the furniture about in a haunted house.

The astronomical evidence for dark matter is now overwhelming, with multiple different types of observations allowing astronomers to map its influence throughout the cosmos. Our best estimates now indicate that there is more than five times more dark matter in the universe than *all* the visible atomic matter, including every star, planet, and speck of dust.

Even more mysterious is a form of repulsive gravity known as dark energy, which is thought to be responsible for causing the expansion of the universe to accelerate. Between them, dark matter and dark energy are thought to make up 95 per cent of the total energy content of the universe. We, and everything we see in the night sky, are just a tiny fraction of a mostly unseen, unknown, and unexplored universe, sparkling froth on the surface of a dark ocean.

There are no particles or quantum fields in the standard model that could be dark matter* or dark energy. This is both a huge challenge for particle physicists and a huge opportunity. While we're highly unlikely to learn about dark energy from particle physics experiments, there is a chance that a dark matter could be found, either in collisions at the LHC or by experiments deep underground that watch for the rare occasions when a dark matter particle bumps into ordinary matter. If we could find such a particle, it could not only explain the motions of stars and galaxies but would give us a clue to whatever larger, more symmetrical picture lies beyond the standard model.

The prospect of creating dark matter was certainly a big motivation for building the LHC, but, surprisingly perhaps, it wasn't the main reason physicists expected to see something new at the collider. There is another mystery, one that has implications that go far beyond just adding another bunch of particles to nature's list of ingredients. It is a mystery that challenges our basic conceptions of the laws of nature and makes us doubt our ability to explain the universe as we find it. It is a problem with the Higgs field, and the strange fact that atoms, people, and apple pies can exist at all.

---

* You might think that neutrinos could fit the bill, but they're too light and zip around the cosmos too fast to match the dark matter data.

LIKE A KITE IN A HURRICANE

About halfway into Monty Python's comedy movie masterpiece *Life of Brian* we find the eponymous hero, Brian of Nazareth, on the run from a troop of Roman centurions. As he scarpers through the streets of first-century Jerusalem, Brian takes a wrong turn up an unfinished spiral staircase, falling with a shriek to what should be his death on the streets far below. However, in a bit of classic Pythonesque surrealism, just before he hits the ground, he falls through the roof of a passing alien spaceship, which is on the run from a second alien craft. After a dramatic chase around the Moon, Brian's ship receives a direct hit from its pursuer and plummets to Earth, crashing at the foot of the same tower that he had just fallen from. As Brian emerges from the smoking wreck unscathed, a bystander who witnessed the whole strange episode exclaims, 'Ooooooh, you lucky bastard!'

'Lucky' is a bit of an understatement. I mean, what are the chances that just at the moment Brian is falling an alien spaceship should just happen to be passing by the Earth, and not just the Earth, but that specific bit of air just above the streets of Jerusalem? Then multiply that bit of amazing luck by the incredible improbability that when the craft is shot down, it crashes back at the very same spot, and what's more that Brian survives the impact. And that's before we get into considerations of the likelihood of intelligent life evolving near enough to the Earth to be able to pop over for a day trip, or the difficult-to-account-for fact that this particular spaceship presumably had a *sunroof*, which its pilots must have absent-mindedly left open.

Yes, 'lucky' doesn't quite cover it. But if we take the standard model at face value, then atoms and therefore anything made of atoms, from stars to human beings, only exist because of an equally ludicrous set of coincidences.

These coincidences relate to the Higgs field, the all-pervading cosmic energy field that is responsible for giving mass to fundamental particles. As we've already discussed, around a trillionth of a second after the big bang, the Higgs field turned on throughout the universe, rising to a nonzero value everywhere. It's this nonzero value that gives fundamental particles mass and basically sets up the ingredients of the universe (and thus our apple pie) as we know them.

With the discovery of the Higgs boson, we know that this field exists, and based on the masses of the W and Z bosons, we can calculate that it settled at a value of around 246 GeV. Now for the important bit. The specific value of the Higgs field is what determines the masses of the fundamental particles. If it helps, you can imagine it like a great cosmic dial, like the kind you use to set the temperature in your house. Reduce it a little bit and the particles of the standard model get lighter; increase it and they get heavier. The trouble is that it seems to be incredibly, fantastically, ridiculously (I'm running out of adverbs here) unlikely that the Higgs field should have settled at what turns out to be a suspiciously perfect Goldilocks value.

Our theories suggest that there should be only two likely values for the Higgs field: 0 GeV or 10,000,000,000,000,000,000 GeV. We'll get to why in a bit, but what you should appreciate first is that either of these two scenarios would be really, really bad if you like existing. If the Higgs field had a value of 0 GeV – that is, switched off – electrons wouldn't have a mass and therefore wouldn't stick to atoms, which alongside a bunch of other bizarre consequences would mean we would not exist. The second scenario, where the Higgs field is turned on all the way, would give fundamental particles such enormous masses that no structures could form without immediately collapsing into black holes. Again, we could not live in such a universe.

On the other hand, 246 GeV is just right to give particles finite but not stupidly big masses, creating a universe full of interesting stuff instead of a haze of massless particles or a load of black holes. However, to get such a pleasant Goldilocks value requires an incredible set of coincidences in the laws of nature, no less improbable than Brian Cohen being saved from death by an alien spaceship.

Ultimately, the origin of this problem lies in the way the Higgs field is affected by empty space, or what physicists call 'the vacuum'. As we've seen, there isn't really such a thing as empty space thanks to the existence of quantum fields, which are always there even when there are no particles sloshing about in them. We saw too that these quantum fields can affect the properties of a particle like an electron – by gathering around it and altering its shape, for example. Well, the quantum fields present in the vacuum should also affect the strength of the Higgs field, and in a catastrophic way.

The source of this catastrophe lies in the fact that even a quantum field with no particles in it is never completely quiet; it is constantly jittering, like the shimmering surface of an almost-still pond. These jitters are due to Heisenberg's famous uncertainty principle, which forbids us from knowing that a field has precisely zero energy. Instead, an empty field must be constantly fluctuating around its zero value.

In principle, these quantum jitters contain energy. How much energy? Well, bizarrely perhaps, it depends on how closely you look at the field. Thanks to the uncertainty principle, as you zoom in on a quantum field and look at it from shorter and shorter distances, the size of these jitters grows larger and larger. This would mean that if you could zoom in infinitely close, the jitters would become infinitely big, giving the vacuum an infinite amount of energy. Fortunately, we know that we can't keep zooming in forever because at a certain extremely short distance, gravity comes into play.

This special distance is known as the Planck length, and it is very, very small: approximately sixteen-trillionths of a trillionth of a trillionth of a metre or, if you prefer lots of zeros in a row, that's 0.000000000000000000000000000000000016 metres. For scale, the Planck length is roughly to a quark what a quark is to you and me; in other words very, very, very small indeed. This distance is special because it is thought that if you could force two particles to within a Planck length of each other, then gravity would cause them to collapse into a tiny black hole. That means that it doesn't make sense to think about distances shorter than a Planck length, and so this is where we stop zooming in.

Even so, since the Planck length is stupendously small, the energy of the jitters in a quantum field at this distance is absolutely enormous. A fairly naive calculation suggests that the energy stored in the quantum jitters of a single quantum field is so large that 1 cubic centimetre of apparently empty space should contain enough energy to blow up every star in the observable universe many, many times over.*

If you are shocked by this result, you should be! Surely that can't be right? The idea that every single sugar-cube-sized bit of space is seething with an apocalyptic quantity of energy seems utterly barmy. Indeed, some physicists doubt the validity of this sort of logic, but it seems to be unavoidable if we accept quantum field theory. Fortunately, this vacuum energy cannot hurt us as it is locked up in space itself and can't be got out. However, even if it can't hurt us, it should have a powerful effect on the Higgs field.

The Higgs field is unique among the fields of the standard model; as we've seen, it is the only one with zero spin. The others are either spin ½ matter fields or spin 1 force fields. This means that unlike the other fields, it feels the effect of these violent vacuum fluctuations like a kite flying in a hurricane.

Imagine releasing a kite in the most powerful hurricane that the world has ever seen. What might you expect to happen? Well, there are probably two likely options: either the wind picks up the kite and drags it high, high into the air or it slams it into the ground,

---

* This is closely connected to another big problem in fundamental physics – the so-called cosmological constant problem. The energy of these vacuum fluctuations should cause space to expand so rapidly that no stars or galaxies could form. Why this terrifying quantity of vacuum energy didn't rip the universe to shreds is one of the greatest mysteries in physics.

pinning it there. It would be extremely surprising if you found the
kite hovering steadily, say, 30 centimetres off the ground.

But this is exactly the situation we find ourselves in with the
Higgs field. Like the kite, the Higgs field's value is buffeted by
these tremendously powerful vacuum fluctuations, which should
either drag it all the way up to the Planck energy (that's
10,000,000,000,000,000,000 GeV) or smack it into the ground at 0
GeV. However, what we find in our universe is that the Higgs field
hovers just above zero at 246 GeV, precisely in the right ballpark
to allow atoms, and therefore the universe as we know it, to exist.

This bizarre situation demands an explanation. The only way to
account for it in the standard model is for the wild fluctuations in
every quantum field we have discovered so far (and even in ones
we haven't discovered) to cancel one another out precisely to an
absolutely unbelievable level of accuracy. This is akin to all the
swirling, howling gusts of wind in a hurricane miraculously
balancing one another so that the air around our kite ends up almost
perfectly calm.

Roughly speaking, the chances of the fluctuations in all the
different quantum fields cancelling one another out to the degree
needed to keep the Higgs field steady at 246 GeV are one in a
million trillion trillion (10^{30}). Such huge numbers are more or less
meaningless, but to put it in some sort of context you'd be significantly more likely to win the lottery jackpot three weeks in a row.

Such an incredible conspiracy between all the different quantum
fields is surely totally implausible. We are left with the impression
that some great cosmic tinkerer has carefully balanced these fluctuations in just the right way to allow atoms to exist. In other words,
the laws of physics look as though they have been fine-tuned for
life.

This smells extremely fishy if you're a physicist, and we're talking
halibut-left-down-the-back-of-the-sofa-for-several-months-over-an-
unusually-hot-summer fishy. This so-called hierarchy problem has
been the greatest motivator for the search for physics beyond the

standard model for the past few decades. The hope is that we can discover some new physical phenomena, be that a new set of quantum fields or something else, that explain why the Higgs field ends up at its perfect Goldilocks value. In our kite analogy, this would be akin to discovering iron rods tethering the kite 30 centimetres above the ground, or perhaps realizing that the hurricane's terrible winds are far weaker than we forecast.

Finding such new phenomena was and remains one of the great goals of the Large Hadron Collider. In fact, along with finding the Higgs, searching for a solution to the hierarchy problem was the main reason the collider was built. The stakes could not be higher. This is more than just another scientific problem; it strikes at the heart of what it means to do physics. Whether we can solve it is intimately bound up with a far deeper issue – namely, whether there are features of our universe that are impossible to explain.

For behind all this lurks a spectre that has haunted physics for decades. Reviled by many, embraced enthusiastically by others, this spectre is the multiverse. This is the idea that our universe is one of a huge, perhaps even an infinite number, with the laws of physics varying from universe to universe. Admit this possibility, and the apparently impossible value of the Higgs field becomes not only probable but inevitable. If we allow for the possibility of other universes, then in the vast majority of them the Higgs field is either at zero or at the Planck scale, and atoms cannot exist. We find ourselves living in a universe where the Higgs field is around 246 GeV not thanks to any miraculous fine-tuning, but because it is the only sort of universe we *can* live in.

If this line of thinking is right, then we will never be able to explain why our universe is the way it is. The Higgs field turned out the way it did by dumb luck, like Brian falling into the path of that spaceship. It was dumb luck that allowed atoms to exist and for life to eventually evolve. What stinks about this way of thinking is that we will never know if it's right or not. There will almost certainly never be a way to detect other universes, as by definition they lie outside our own and out of reach.

To put it another way, if the multiverse is right, then we can never know how to make an apple pie from scratch.

What many physicists hope, though, is that some unknown effects are responsible for stabilizing the Higgs field against catastrophe. If this is right, then there are good reasons to believe that new particles should exist, with masses similar to the Higgs boson itself. So just as the Higgs was emerging from the data at the Large Hadron Collider, hundreds of other physicists were scouring the collisions in search of a chink in the standard model's armour that might explain why we live in such an impossibly unlikely universe.

INTO THE UNKNOWN

Every Wednesday morning, a group of physicists squeeze into a windowless meeting room on the first floor of the Cavendish Laboratory in Cambridge. Sitting around a large table scattered with coffee cups and lit dimly by a frosted skylight, the discussion is animated and peppered with strange vocabulary. 'Squarks', 'neutralinos', 'gravitons', 'Z primes', and 'micro black holes' fly back and forth across the table. Every so often someone will jump up to scribble something on a whiteboard, hieroglyphs of arrows and wiggles or semidecipherable scrawls of mathematical symbols, while others argue from their seats, watch on pensively, or tap away on their laptops.

The Supersymmetry Working Group has been meeting since before I arrived at the Cavendish in 2008. While there are a hell of a lot of meetings in particle physics,* what makes this one unusual

---

* Urban legend holds that the ATLAS experiment once set up a group to look at ways of reducing the number of meetings, which only exacerbated the problem when it started holding regular meetings.

is that it is a coming together of experimentalists working on the LHC along with theoretical physicists from the Cavendish and the maths department down the road. For more than a decade now, they have trawled through recent results from the LHC and the latest theoretical ideas in pursuit of new schemes to search for exotic phenomena.

Among the regulars are Ben Allanach and Sarah Williams. Ben, a professor of theoretical physics, has spent the past ten years helping to guide experimentalists down promising avenues while figuring out what the latest LHC result means for speculative theories that go beyond the standard model. Meanwhile, Sarah has been on the hunt for signs of something new in the trillions of collisions recorded by the ATLAS experiment.

For years, by far the most promising of these speculative theories was supersymmetry, an idea so seductive that Ben has spent his entire career thinking about it, while Sarah and hundreds of her collaborators on ATLAS have performed dozens upon dozens of measurements in the hope of spotting its effects.

Supersymmetry is the rarest of ideas, one that solves several deep, fundamental problems in a single stroke. It promises to explain how matter gained dominance over antimatter during the big bang, the nature of dark matter, and even suggests that all the forces of nature were once unified in the very earliest moments of our universe. However, perhaps its greatest appeal is that it protects the Higgs field from the violence of the vacuum, naturally explaining why its strength is set at just the right value to allow atoms to exist.

As the name suggests, supersymmetry imposes a new symmetry on nature's fundamental building blocks, not all that dissimilar to the symmetry relating matter to antimatter. However, instead of relating particles to their antiparticles, supersymmetry relates matter particles like electrons, quarks, and neutrinos to force particles like photons, gluons, and Higgs bosons.

What distinguishes a matter particle from a force particle is its spin. All matter particles are fermions with spin ½, while the force particles are bosons with spin 1, or 0 in the special case of the Higgs. According to supersymmetry, for every spin ½ matter particle in the standard model there should be a spin 0 'superpartner'; for every force particle there's a spin ½ superversion. These superparticles have identical properties to their standard model partners, only differing in their spins.

These supersymmetric particles all have very silly names; the superversion of the electron is called the 'selectron', while the partners of the quarks are called 'squarks'. It's not any better for the supersymmetric versions of the force particles; the photon's partner is the photino, and there are gluinos, winos, zinos, and higgsinos. Perhaps my least favourite is the sstrange squark, which when said out loud usually makes you inadvertently spit in your colleague's eye. Together, these supersymmetric particles are called 'sparticles'. Part of me hopes that supersymmetry is never found just so we don't have to use these silly words ever again.

Clunky nomenclature aside, supersymmetry is regarded by many theorists as one of the most beautiful and powerful ideas to have been discovered in fundamental physics. In particular, it is one of the few ways that theorists have found to save the Higgs field from catastrophe. As we've discussed, the Higgs field is uniquely sensitive to fluctuations in the quantum fields that are ever present in the vacuum. Each of the twenty-five or so quantum fields in the standard model contributes its own set of fluctuations, each acting like a hurricane-force wind that should blow the Higgs field down to zero or up to the Planck energy. There is absolutely no reason to expect that all these different quantum winds should balance one another out, which is why the Higgs field hovering stably at 246 GeV is so hard to understand.

Supersymmetry solves this problem. For every quantum field in the standard model there is now a corresponding superfield, and when you study the mathematics you find that the fluctuations in a superfield are almost exactly equal and opposite to the fluctuations in its standard model partner. So, for instance, when the electron field blows the Higgs one way, the selectron (superelectron) field blows it back in the opposite direction, like two countervailing winds that almost completely cancel each other out and turn what was a quantum mechanical hurricane into the equivalent of a clear, calm day.

With supersymmetry, you no longer need to resort to fine-tuning or untestable multiverses. The theory is *natural*, which means it automatically explains why the world is as it is without the need to fiddle about excessively with the theory. Better still, in many versions of supersymmetry the lightest sparticle is a perfect candidate for dark matter.

However, there is an obvious objection: where are all the sparticles? If the universe is perfectly supersymmetric, then apart from having different spins, the sparticles should have exactly the same properties as their standard model buddies, including the same masses, and if that were true, we would have discovered them already. To get around this, supersymmetry has to be imperfect, like one of those warped fairground mirrors that make you look like you've been run over by a steamroller. Breaking supersymmetry allows the sparticles to be heavier than the ordinary standard model particles, heavy enough that previous colliders wouldn't have had enough energy to make them and so explaining why we haven't seen them yet – but this comes at a price. The more you break supersymmetry by making the sparticles heavier, the less effective it is at cancelling out those nasty quantum fluctuations. The upshot of all this is that if supersymmetry is to save the Higgs, then sparticles can't be much heavier than the Higgs itself. That would put them squarely in the sights of the LHC.

Given its huge promise, it's unsurprising that the lure of supersymmetry has been impossible to resist, for theorists and experimenters alike. However, it isn't the only show in town when it comes to stabilizing the Higgs field. While supersymmetry saves the Higgs by calming the quantum hurricane with a superstorm of equal and opposite strength, another popular set of approaches argues that there was never a hurricane in the first place.

The enormous vacuum fluctuations that are so dangerous to the Higgs are a consequence of the fact that as we zoom in on the vacuum to ever-shorter distances, the fluctuations appear to grow larger and larger. As we've seen, this zooming-in process goes on all the way down to the Planck length, the point at which two particles forced together collapse into a black hole.

The Planck length is short ultimately because gravity is an incredibly weak force, a trillion trillion trillion times weaker than electromagnetism. This means that in any particle physics experiment we can currently perform, it is totally overwhelmed by the other three quantum forces. It would only start to match them in strength if you could get two particles incredibly close together, which would in turn mean firing them at each other with an enormous amount of energy. At the LHC we have enough energy to get down to about 10^{-18} metres, which is darn small but still a hundred thousand trillion times bigger than the Planck length where gravity becomes strong.

But what if gravity is actually stronger than it appears? If that were true, then the point at which you collapse two particles into a black hole would happen earlier, which means that we'd stop zooming in on the vacuum earlier too. And if you stop zooming in earlier, then the size of the quantum fluctuations is also much smaller, effectively reducing what was a hurricane-force wind to a gentle quantum breeze.

The way you do this – and bear with me here – is by introducing extra dimensions of space. We live in a 3D world, where we can move forwards and backwards, up and down, and left and right, but in these extradimensional theories there are new directions that you can move in. I invite you to imagine what moving in, say, a fourth dimension would be like. No? Me neither. Since our brains evolved to navigate a 3D world, it's impossible for us to visualize higher dimensions (and if you ever hear a mathematician or physicist telling you they can picture a 4D world I'm pretty sure they're lying or high). But mathematically at least, these ideas are straightforward to write down. In such theories, the reason that we don't perceive these extra dimensions is usually because they are incredibly tiny or because the particles we are made from are stuck in our 3D world, like stickmen drawn on a sheet of paper.

Gravity, on the other hand, *is* able to access these higher dimensions, allowing it to leak away like water from a dodgy pipe. This leaking explains why gravity appears weak in our ordinary 3D world, whereas if we could perceive all the dimensions, we'd realize that gravity is just as strong as the other forces.

This may all sound like sci-fi speculation, but what is neat about these extradimensional theories is that, like supersymmetry, they predict new phenomena that should appear at the LHC. If these extra dimensions exist, then the energies required to make tiny black holes is far lower than the Planck scale, making it possible to create them in the collisions at the LHC after all.

The prospect of making microscopic black holes led to a now-familiar round of apocalyptic headlines, particularly in the British tabloid press. Just before the LHC switched on in 2008 the *Daily Mail* published a story with the typically level-headed headline 'ARE WE ALL GOING TO DIE NEXT WEDNESDAY?' while in the United States, *Time* magazine went with the slightly less alarmist but nonetheless startling 'COLLIDER TRIGGERS END-OF-WORLD FEARS'. These fears were that once made, a tiny black hole would sink to the centre of the Earth and slowly devour the entire planet.

Eventually, the media frenzy became so great that CERN assembled a panel of experts to assess the various doomsday scenarios. They produced a truly excellent document titled *Review of the Safety of LHC Collisions*, which is probably the most exciting risk assessment ever written and worth a read just for lines like 'The possible concern about high-energy particle collisions is that they might stimulate the production of small "bubbles" . . . which would then expand and destroy not just the Earth, but potentially the entire Universe.'

Exciting stuff. Fortunately, the panel concluded that since the universe has been steadily bombarding the Earth with cosmic rays far higher in energy than we can achieve at the LHC, if such doomsday events were possible they would already have happened and the Earth and every other heavenly body would have been destroyed long ago. The panel appears to have been right as the world still seems to exist, so far at least. Anyway, I guess if the world does end, no one will have time to sue.

The reason that tiny black holes aren't thought to be a threat is encapsulated in Stephen Hawking's famous prediction that they should evaporate by emitting Hawking radiation. For big, star-sized black holes that lurk in deep space, this process is absurdly slow, but tiny black holes of the size that might be made at the LHC would disintegrate almost instantly into a spray of particles, which could then be spotted by the giant ATLAS and CMS detectors.

Existential terror aside, supersymmetry and extra dimensions remain the two most popular ways to explain the strength of the Higgs field, though they are by no means the only ones. Regardless of which phenomenon is ultimately responsible for keeping our metaphorical kite aloft, you would more or less always expect to find something new close in energy to the Higgs boson itself. So when the LHC started colliding in 2010, hopes were high that we would soon glimpse new ingredients of our universe along with the Higgs itself.

The first year of the LHC's running was basically a warm-up lap, as the engineers driving the collider in the CERN Control Centre learned how to operate their shiny new machine. When collisions restarted in the spring of 2011 after a winter break, the collider raced out of the starting blocks, accumulating more data in the space of a few days than in the whole of the previous year. Now the game was well and truly on.

As Christmas 2011 approached, telltale hints of the Higgs boson were already emerging from the data recorded by ATLAS and CMS. However, despite hopes that sparticles might reveal themselves at the same time, all the searches were drawing blanks. Still it was early days for the LHC, nothing to get too concerned about just yet.

Fast-forward to July 2012; CERN jubilantly announced the discovery of the Higgs boson to a world that briefly became fascinated by particle physics. However, as champagne bottles were uncorked in the offices of CERN, anxiety was already growing about the absence of any of the *other* predicted new particles. My colleague Sarah Williams, newly arrived at CERN as a PhD student and sleep deprived after weeks of intense work, had just unblinded a search for supersymmetric versions of the leptons, or sleptons (I know, right?). Despite huge excitement among her more senior colleagues that they were about to see something new, when they looked at the data there wasn't even a whiff of a sparticle.

In fact, every search for supersymmetry, micro black holes, and other exotica had come up empty-handed. Perhaps even more troubling was the mass of the newly discovered Higgs boson. The simplest supersymmetric theories generally predicted a Higgs with a mass close to the Z boson, around 90 GeV; however, the particle seen by ATLAS and CMS was uncomfortably heavy, up at 125 GeV. Although this could be accommodated with a bit of theoretical gerrymandering, it started to put the theory under strain.

Towards the end of 2012, my own experiment, LHCb, announced more bad news for supersymmetry fans. We had found evidence for an extremely rare decay of the beauty quark that was predicted to get a big boost in certain versions of supersymmetry. However, the decay had been seen with a rate that was in more or less perfect agreement with the standard model. When the BBC broke the story it ruffled a few feathers, particularly thanks to a quote from my colleague Chris Parkes at the University of Manchester, who declared that the new result had put supersymmetry 'into hospital'. My boss and head of the Cambridge LHCb group, Val Gibson, joined the fray, saying that the result was 'putting our supersymmetry theory colleagues in a spin'. After all, experimental physicists like nothing better than proving their smarty-pants theory colleagues wrong. At CERN, the eminent theoretician John Ellis, who had spent more than thirty years working on supersymmetry, fired back dismissively that the result 'was actually expected in (some) supersymmetric models. I certainly won't lose any sleep over the result.'

How can a bunch of respected professors of physics have such different interpretations of a result? Well, an important point to understand about supersymmetry is that it is not a single theory – it is a principle that can be used to build a huge number of different theories with wildly different predictions. The upshot of this is that supersymmetry is fiendishly difficult to kill. If your favourite supersymmetric model doesn't show up at the LHC, it's almost always possible to adjust some of its parameters or add some extra bells and whistles to explain why you haven't seen it yet. However, as you tinker and augment to explain away its failures, you begin to corrupt supersymmetry's very purpose. After all, it was invented to avoid fine-tuning in the standard model, so fine-tuning supersymmetry itself feels like a betrayal of its founding principle.

The final protons of the first run of the LHC smashed into one another just before Christmas 2012. As the engineers who had built and operated their remarkable machine looked back on the past three years with justified pride, the physics community struggled to make sense of the landscape that the LHC had revealed. Against great expectations of a rich vista full of new and exciting opportunities for exploration, the LHC had revealed a wasteland, at the centre of which stood the solitary Higgs boson, an inexplicable tree alone in an arid desert.

Some physicists began to whisper of the 'nightmare scenario' – the possibility that the LHC would ultimately only discover the Higgs while providing no other clues to the great problems of fundamental physics. Some young people began to reassess their career plans. Matt Kenzie, who had been involved in the Higgs discovery at CMS, made the bold decision to switch to LHCb after his PhD, believing that the writing was already on the wall for the chances of seeing new particles at ATLAS or CMS. Older heads advised caution. It was still very early days, they said. We had waited more than thirty years for supersymmetry; we could afford to wait a little longer. Ben Allanach at Cambridge captured the mood of many of his theory colleagues: 'Supersymmetry is a bit late to the party, but I don't think it's lost yet.'

Hope was on the horizon. After two years of engineering work to remedy the faults that had forced the LHC to run at around half its maximum energy for the first few years, the collider restarted with a record-breaking collision energy of 13 TeV in May 2015. Once again, the frontier of exploration was being pushed further into uncharted territory. Perhaps the promised riches were finally within reach.

Then, just before Christmas 2015, a bolt from the blue. ATLAS and CMS released evidence of a new bump appearing in the high-energy data recorded that year. In eerie echoes of the Higgs hints that had been revealed before Christmas 2011, they were both seeing evidence of a new particle decaying into two photons, but six times heavier than the Higgs, with a mass all the way up at 750 GeV.

The accumulated tension that had been building in the theory community for more than five years was suddenly released in a torrent of speculative proposals explaining the new bump. Within a few weeks more than five hundred papers had been uploaded to the online preprint repository,* including one from Ben and his colleagues. Many speculated that the new discovery might be one

---

* The site is arXiv.org – an online repository where scientific articles are uploaded before they've been peer-reviewed or published in a scientific journal.

of the long-awaited superparticles, the herald of a new quantum army that would soon march into view.

In August the following summer, physicists gathered in Chicago for the biggest particle physics event of the year, the International Conference on High Energy Physics. ATLAS and CMS were both scheduled to give eagerly anticipated updates on the 750 GeV bump using the additional data that had been recorded that year. However, the night before the talks, CMS jumped the gun by accidentally posting their paper online. The result landed like a punch to the gut. As more collisions had accumulated, the bump had melted away to nothing. It appeared to have been little more than a random fluctuation in the data. More than five hundred papers had been written about a cruel statistical fluke.

Meanwhile, Sarah was working as part of a team on ATLAS searching for signs of micro black holes using the 2015 data. There were hopes that with the higher-energy collisions it might now be possible to create them, but again, they failed to show.

The LHC and its giant experiments continued to perform admirably for a further three years, churning out vast quantities of high-quality data, smashing almost all the expectations set at the start of the run. By the time the machine switched off again on 3 December 2018 for another scheduled two-year shutdown, the LHC experiments had recorded more than ten thousand trillion collisions, and yet not a single hint of a new particle beyond the Higgs had been spotted among all the subnuclear debris. The nightmare scenario seems to be coming true.

Fundamental physics now faces a crisis unlike anything seen in the last one hundred years. We know that there are major features of our universe that we do not understand: the origin of matter during the big bang, what dark matter is made from, and, most of all, how we find ourselves living in a universe that appears spookily fine-tuned for life. And yet, the machine that was built to provide us with answers, the largest ever built by humankind, has offered up only the same standard model that we know must be incomplete. This isn't a failure of the experiment; the LHC is an engineering and technical triumph. It has simply shown us how nature is, and nature, it seems, doesn't care much for our clever theories.

Although many still hold out hope that supersymmetry will appear at the LHC in the coming years and save us from the pseudo-scientific jaws of the multiverse, others have refocused their efforts in what might be more fruitful directions. Supersymmetry – at least in the grandest and most ambitious forms that explain the strength of the Higgs field, the nature of dark matter, and the unification of the forces all in one go – appears to have failed. To have escaped detection, the predicted sparticles must now be so massive that they would no longer balance the powerful vacuum fluctuations that threaten to blast the Higgs field away from its Goldilocks value, leaving us with an uninhabitable universe. Seeing the way the wind was blowing, at the start of 2019, the Cambridge Supersymmetry Working Group quietly rebranded itself as the Phenomenology Working Group.

So where do we go from here? Is this it, the end of the road? Are there simply features of the universe that are beyond our power to explain? It may be a cliché, but every crisis is an opportunity, and this crisis presents a huge one. The LHC may not have given the answers we were hoping for yet, but it is telling us *something*. The challenge now is to figure out what. This is a moment to re-examine our assumptions and look at old problems from a different angle. More than anything, it is a time to put our grand ideas and preconceptions to one side and to listen carefully to what nature is saying.

In fact, it may already be speaking to us in unexpected ways. Over the past few years, a series of strange and unanticipated signals have started to emerge from the LHCb experiment, signs, at long, long last, of nature deviating from the standard model. It's too early to be certain just yet, but maybe, just maybe, we are about to see a deeper layer of the cosmic onion.

THE AGE OF ANOMALIES

LHCb doesn't usually get the same attention as the big beasts, ATLAS and CMS. We didn't discover the Higgs (to be fair, we didn't look for it), nor do we search for sexy-sounding stuff like dark matter or micro black holes. And compared to the photogenic ATLAS and CMS detectors, which look like alien portals to another dimension, go down to the LHCb cavern and you'll be confronted with something that looks kind of like a giant multicoloured toast rack.

However, as ATLAS and CMS have been burning through speculative new theory after speculative new theory, LHCb has emerged as the best hope for finally seeing something beyond the standard model at the LHC. For the past few years, anomalies have started to appear in the data, anomalies that may be hinting at something altogether new.

To understand the difference in approach between ATLAS, CMS, and LHCb, consider two hunters standing at the edge of a dense jungle. Somewhere, out there, in the miles and miles of tangled foliage, is an elephant. Or so they have been told by a local elephant theorist. One hunter strides confidently into the undergrowth, hacking his way through vines and bracken, pushing ever deeper into the forest in search of his quarry. But the jungle is big and dark and grows thicker and more oppressive with each step, and he eventually reaches a point where he can't go any farther without having caught sight of the elephant.

Meanwhile, his companion has wandered in only a little way, where shafts of sunlight still pierce the canopy and the going is a little easier. She moves slowly and methodically, her eyes scanning the forest floor for something out of place – a footprint, or perhaps a broken branch. After a long while, she notices a faint depression in the soft earth, as wide as a tree trunk with four marks that might be toes. A little later, she finds another, and then another, leading her deeper and deeper into the jungle. The elephant is out there. She is on its trail.

The hunter hacking his way through the jungle is like ATLAS and CMS, the two giant general-purpose detectors that scour through trillions upon trillions of collisions in search of new particles hidden in the quantum undergrowth. This sort of direct search can work well if you have a clearly defined target and know what energy range/bit of the jungle to search in. It was exactly how the Higgs was found, for instance. However, if the particles you are looking for lie out of reach – perhaps they're too heavy to be produced directly in your collisions or are especially well camouflaged among the ordinary particles – then you'll draw a blank.

But there is another approach, a so-called indirect search. Like the hunter scanning the ground for footprints, it's possible to detect hints of new quantum fields through their effects on ordinary standard model particles. The advantage is that you can spy evidence of a new quantum field even if the associated particle is too massive to create directly. The disadvantage is that you may not be able to figure out precisely what is causing the effect, like a hunter who can't quite tell what species of elephant they're tracking just from its footprints.

Broadly speaking, this second, indirect approach is the one we take at LHCb. Unlike the multipurpose ATLAS and CMS experiments, LHCb is specifically designed to study standard model particles with high precision in the hope of catching them misbehaving. As I've noted, the 'b' in the LHCb stands for 'beauty', the heaviest cousin of the ordinary down quark that makes up atomic matter. This negatively charged quark is more usually referred to as the 'bottom quark'; there was an attempt to name the two heaviest quarks 'truth' and 'beauty', but the community plumped for the less poetic 'top' and 'bottom'. At LHCb we'd rather be known as beauty physicists than bottom physicists, and so, for us at least, it's beauty not bottom.

The beauty quark is interesting because it's particularly sensitive to the existence of new quantum fields, which can do things like alter how long it lives before decaying, or how often it decays into different sorts of particles. One of the best ways to try to spot these sorts of effects is to study decays of the beauty quark that are predicted to be extremely rare in the standard model.

Take, for instance, a beauty quark decaying into a strange quark, a muon, and an antimuon. There is no simple way to make this decay happen in the standard model; the decay has to go via a complicated mixture of different quantum fields, including the W and Z bosons and the top quark field. It's a bit like trying to get between two London Underground stations where there's no direct route, forcing you to change trains several times. Most people would avoid the hassle of such a convoluted journey; as a result, the number of passengers travelling between the two stations is very low. Likewise, the fact that our beauty quark decay involves so many different quantum fields makes it very rare indeed.

But what if there were a more direct route, one that, for the sake of our analogy, doesn't use the normal underground network? For instance, maybe there's an overland train that goes directly between the two stations in a little over twenty minutes. The same could be true for our beauty quark if a new quantum field, for instance a new force of nature, exists that provides a more direct way for it to decay. This can even happen if the particle of the new field is far too heavy to be created by the LHC. Even without having particles moving about in the field, the field is still there, and some energy can still go through it briefly without actually having to create the associated particle.*

So if we count how often a beauty quark decays into a strange quark, a muon, and an antimuon and compare that to what the standard model predicts, then we can potentially detect the influence of unseen, undiscovered quantum fields. These decays are

---

* Exactly the same thing occurs when a neutron decays into a proton. This happens via the W boson field, even though the W boson particle is more than eighty times heavier than a neutron and therefore way too heavy to be created directly in the decay.

extremely rare – only one in a million beauty quarks will decay in this way – so to have any chance of spotting it, you need to make a hell of a lot of them.

Fortunately, the LHC is brilliant at making beauty quarks. Since protons are made of quarks bound together by the gluon fields, when you smack them into each other you tend to get lots of quarks. In a year, the LHC will create billions of beauty quarks and anti-quarks inside LHCb, whose design has been specifically honed to study them.

Since these decays are really, really rare, it took a while for LHCb to collect enough collisions to be able to make precise enough measurements. However, every year, billions more beauty quarks were produced, and more and more of these rare decays were spotted. At first, the results all seemed to agree with the standard model fairly well, but as the precision improved, hints of small deviations started to appear.

The first big clue came in 2014, when a team at LHCb compared how often a beauty quark decays into a strange quark, a muon, and an antimuon to the equivalent decay but swapping the muons for electrons. As far as the forces of the standard model are concerned, the electron and its heavier cousins the muon and the tau are completely identical, the only difference being that the muon is 200 times heavier and the tau a whopping 3,500 times heavier than the electron. The fact that the forces treat these three leptons the same is known as 'lepton universality', and it's a hard rule of the standard model. Lepton universality means you'd expect the beauty quark to decay to muons just as often as it decays to electrons.

But that isn't what the team found.

Instead, it seemed that the muon decay was only happening 75 per cent as often as the electron one, as if beauty particles *preferred* to decay into electrons. That said, the uncertainty on the measurement was pretty big, about 10 per cent, so there was a good chance that it was just a random fluctuation like the bump that had fooled everyone at ATLAS and CMS in 2015. However, a couple of years later another measurement using a separate data sample found a very similar effect. This time, the muon decay was only happening around 69 per cent as often as the electron one, and what's more, the uncertainty was smaller.

It was at this point that the theory community started paying attention. As signs of new particles faded away at ATLAS and CMS, something seemed to be emerging from the data at LHCb. Further measurements of different beauty decays involving taus (the heaviest copy of the electron) saw similar effects. Meanwhile, thousands of miles away, the BaBar experiment in California and the Belle experiment in Japan had also seen hints of beauty decays breaking the sacred rule of lepton universality. On their own, none of these deviations are significant enough to declare that the standard model had finally been broken, but as more and more anomalies have cropped up, a coherent picture could be starting to emerge.

At the start of spring 2019, I met up with theorist Ben Allanach in his office at the Department of Applied Mathematics and Theoretical Physics. Ben is a supersymmetry expert, having spent his entire career working on various models and helping his experimental colleagues come up with new ways of searching for sparticles at the LHC. However, the slew of negative results from ATLAS and CMS has turned him away from the subject, for now at least.

'A lot of people got a little depressed, particularly those of us who did supersymmetry for a long time. There were a plethora of different reactions and some people are still going strong, but I think a lot of people have got turned off it.'

As far as Ben is concerned, the anomalies in beauty decays are where the action is: 'They're the best hope we have at the moment, I definitely think it's exciting.' The big question on everyone's minds now is, are these anomalies the real deal? After all, we've been fooled by unlucky statistical flukes before. Ben doesn't think that's likely in this case: 'There are just too many of them for it to be a fluctuation. There's something going on.' The bigger fear is that these anomalies could be due to some misunderstood effect, either in the theory of how quarks behave or something we've got wrong in our experiments. While huge care is taken to try to account for every possible effect that might bias your result, these big particle detectors are unbelievably complicated and there's always a chance that you could miss something.

'If you were a betting man,' I asked, 'where would you put your money?'

Ben paused for a moment and turned to look out of the window. 'Well, you'd really want another experiment to independently confirm it . . .'

'But if I pushed you.'

'I'd take just about evens on it being real new physics – which is really high. It's the best thing I've seen in my career.'

Since moving away from supersymmetry, Ben has started taking a different approach to solving problems. Instead of working on grand theories based on a single elegant principle that solves lots of problems at once, he's now listening carefully to what the data are saying and trying to build understanding from the bottom up. So if these anomalies are real, and that's a big if, what could be causing them?

'There are basically two camps. Either it's something called a Z prime – a new force field – or a leptoquark.' These are essentially new quantum fields that are interfering with how the beauty quarks decay. A Z prime would be a force field much like the Z boson of the weak force, except one that breaks lepton universality, for instance by pulling more strongly on muons than electrons. A leptoquark, on the other hand, would be a rather more exotic beast.

One of the great mysteries of the standard model is why it contains twelve matter particles – six quarks and six leptons – and why they appear in three copies or generations. The electron, up quark, and down quark that make up our apple pie sit in the first generation, with additional, heavier, unstable copies of these particles in the second and third generations. The patterns in these matter particles echo the patterns in the periodic table of the chemical elements that Mendeleev drew up in the nineteenth century. In the case of the chemical elements, those patterns pointed to a deeper structure, which was eventually revealed to be the quantum structure of atoms. Could the matter particles of the standard model be hinting at something similar?

A leptoquark would be a new particle that could decay to both leptons and quarks at the same time, acting as bridge between these two different and apparently unrelated types of matter particles. If such a thing existed, it could be the first part of a jigsaw puzzle that could eventually reveal the ultimate origins of the matter particles that make up our universe.

That would be a *huge* deal, arguably the biggest discovery in particle physics since the standard model was first written down. When the anomalies started to build up, Ben and his colleagues began conservatively by just adding an extra quantum field to the standard model to see if you could explain all the anomalies at the same time. Now they are starting on the harder project of trying to figure out whether these new quantum fields fit into a bigger, more elegant structure.

Despite the LHC finding no evidence of supersymmetry, Ben still thinks that something *must* explain the fine-tuning of the Higgs field. 'It's like dropping a pencil on a table and it landing exactly nib down and standing upright. The fact that it's staying up is telling us something.' Amazingly, one of the theories they've been studying to explain the beauty anomalies may also do the job that supersymmetry has failed to do, namely to stabilize the Higgs field and prevent the universe from collapsing into an uninhabitable wasteland.

The explanation for both effects may be that the Higgs boson is not a fundamental particle but a mixture of other new fundamental quantum fields. The reason the Higgs was thought to be so sensitive to fluctuations in the vacuum, like the kite in the hurricane, is that it has zero spin. However, if it's made of other fields stuck together whose spins just add up to zero, then it would no longer be affected by those nasty vacuum fluctuations. What's more, the new quantum fields that make up the Higgs might also explain the patterns in the matter particles of the standard model.

We're at a turning point in our understanding of the ingredients of the universe, a moment of anxiety and crisis, excitement and opportunity. No one knows whether these anomalies are real, whether they'll get stronger or gradually fade away. But whatever happens, nature is speaking to us. Of course, we all hope that these anomalies are real, because if they are, we will have finally peeled back another layer of reality and seen the first signs of what lies beyond the standard model. It would also be great for experimental physicists like me, the beginning of a new era of exploration that could be even more exciting than the heady days of the 1960s and 1970s when the basic building blocks of nature were being discovered.

But if the worst happens and these anomalies melt away, we will nonetheless have learned something profound. If in 2035, when the LHC powers down for good, we have still found nothing apart from the Higgs and the nightmare scenario is realized in its full horror, it may be the crisis needed to trigger a rethink of our approach to fundamental physics. It will be clear that we do not understand something deep about the nature of quantum fields, the vacuum, possibly gravity too. Because if we want to go right back to the moment the universe began, to the 'b' of the big bang, we are going to need a complete picture that describes all three together. It is only then, as Carl Sagan put it, that we'll be able to invent the universe.

CHAPTER 13

Invent the Universe

I t's time to face facts: we are still a long way from knowing how to make an apple pie from scratch. Although there are lots of promising ideas out there and we are continually learning more from experiments and observations, we don't yet know how the particles in our apple pie ultimately survived the big bang, and we can't explain why the Higgs field settled at just the weirdly specific value that makes the existence of atoms possible. We don't know what dark matter is, and without dark matter's gravitational influence ordinary matter would never have clumped together in large enough quantities to form galaxies, stars, and planets, and you need planets and stars to grow apples.

Even putting these mysteries aside, we aren't sure whether we might be missing other quantum fields beyond the ones in the standard model. We can't even really explain why our universe contains the quantum fields it does, or whether the quantum fields we do know about might be made of even more fundamental ingredients. And these are just a handful of the questions that we know we don't have answers to – or, to borrow a phrase from former US Secretary of Defense Donald Rumsfeld, the 'known unknowns'. There are almost certainly a whole load of unknown unknowns as well, questions that are so far beyond our ken that we haven't even thought to ask them yet. In other words, we have a hell of a lot still to learn.

So we don't yet know how to make an apple pie from scratch, but perhaps an even bigger question is, will we ever be able to find out? Over the course of this book, we've seen how thousands of women and men, chemists, physicists, and astronomers, experimenters and theorists, technicians and machine builders, engineers and computer scientists, working together over hundreds of years, have gradually dismantled matter into its most basic components and tracked their origins out into the cosmos, through the hearts of dying stars and eventually all the way back to a trillionth of a second after the big bang. The fact that we can tell so much of this story is one of humankind's greatest achievements. The question is, how much further does this story go, and can we ever really discover a complete description of how the universe began?

To try to make this question a little more concrete, let's first consider what the ultimate apple pie recipe would need to be like in order to qualify as starting 'from scratch'. To explain where the matter in our apple pie ultimately came from, we would need a theory that could describe what happened at time zero, the moment the universe began, or as Carl Sagan put it, we need a theory that invents the universe.

Modern fundamental physics stands on two theoretical pillars: quantum field theory describing the microworld of atoms and particles; and general relativity, the theory of the force of gravity, which sculpts the universe at vast scales. While dazzlingly successful in their own domains – and to be clear no experiment or observation is in conflict with either theory – it is clear that both will fail us as we approach the moment of the big bang.

The reason is actually pretty simple when you come down to it: quantum field theory ignores gravity and general relativity ignores quantum mechanics. This is absolutely fine for almost every situation that either theory might normally be asked to explain. On the one hand, since gravity is a trillion trillion trillion times weaker than electromagnetism, when you're doing experiments down at the level of particles, the gravitational forces are utterly negligible compared to the much more powerful influence of the three quantum forces. On the other hand, if you're an astrophysicist or cosmologist working at the scale of stars, galaxies, or the entire universe, then (except in one very important case that we'll get to shortly) there's no need to bother about piddly little quantum effects down at the subatomic level.

However, at the moment of the big bang, the whole cosmos was subatomic. Literally everything – energy and fields, space and time – was compressed into an infinitesimally tiny point, far, far smaller than an atom. Under these unimaginably extreme conditions both gravity and quantum mechanics would have ruled the universe *together*. To describe this first moment, particle physics and cosmology, quantum field theory and general relativity, must merge into a unified quantum theory of gravity.

Finding a quantum theory of gravity has been regarded as the holy grail of theoretical physics for almost a century. Generations of physicists have worked at the problem, and while several potential candidate theories have been found – string theory, loop quantum gravity, causal dynamical triangulations, and asymptotically safe gravity, to name a few – no one knows whether any of them actually describe the real world.

Nonetheless, if we could find such a theory, we would at least have the language needed to describe the moment the universe began. But in its most ambitious form, the ultimate recipe would go even further than this. Not only would it be a quantum theory of gravity and be able to describe the birth of the universe, it would also explain why the universe contains the basic set of ingredients it does and why they are the way they are. It would explain, for example, why there are six quarks and six leptons, why they have the masses and charges they do, and why there are three quantum forces and why they are as strong as they are. It would explain the strength of the Higgs field, what dark matter is, and how matter got made in the big bang. In other words, it would be what physicists often call a 'theory of everything'.

This is the kind of superambitious ultimate theory that Steven Weinberg, one of the architects of the standard model, described in his book *Dreams of a Final Theory*, published in 1992. Weinberg's vision was of a theory founded on a principle of such beauty and power that it would account for all the apparently arbitrary features of the quantum world without the need to put anything in by hand. This theory would be unique, elegant, and so rigid that any attempts to modify it would make the whole thing collapse. It would in some sense be inevitable, a final explanation that needs no further explaining.

Now that is an incredibly high bar to reach. Nonetheless, when Weinberg wrote *Dreams* in the early 1990s, there was a sense among some theorists that such a theory was starting to reveal itself – or as Weinberg put it, 'We think we are beginning to catch glimpses of the outlines of a final theory.'

Weinberg was referring to string theory, which for the past forty years has been by far and away the most popular approach to finding a theory of quantum gravity, and which seemed like it might also be the unique theory of everything.

THE ULTIMATE THEORY

On a leafy suburban street on the outskirts of Princeton, New Jersey, stands what appears from the road to be a relatively modest white clapboard house with a small neatly kept garden at the front. A painted wooden sign leaning against the steps to the porch warns in large weary letters that this is a 'Private Residence', in what I suspect is a futile attempt to keep curious tourists from peering in through the windows.

This is the house where Albert Einstein lived for the final twenty years of his life, having left Germany for the last time in 1932 to escape persecution by the Nazis. Visitors to 112 Mercer Street would usually find the ageing, wild-haired Einstein in his study, dressed comfortably in a soft sweater and surrounded by papers covered in algebraic symbols. George Gamow, who was an occasional caller in the late 1940s, recalled glimpsing these papers during their conversations, and though Einstein seemed as sharp as ever, he would never bring up what he was working on.

Einstein had been catapulted to worldwide fame when his general theory of relativity had been spectacularly confirmed by measurements made during a total solar eclipse in 1919. With general relativity, Einstein had radically reimagined the concepts of space, time, and gravity, superseding the work of the man widely regarded as the greatest physicist in history, Isaac Newton. According to Einstein, space and time weren't merely coordinates that tell you when or where an event takes place but a physical fabric that could be bent, stretched, compressed, and even set vibrating, like the elasticated surface of a trampoline. Newton had never been able to explain what gravity was, and when confronted with the question of how the Earth reaches out across empty space and pulls on the Moon, had famously written, 'I feign no hypothesis.' Einstein resolved the conundrum, showing that the force of gravity is an illusion. Instead, the Earth bends spacetime around it, like a bowling ball resting on said trampoline, and the Moon merely follows the closest thing to a straight line (technically what is called a geodesic). It's just that close to the Earth, straight lines are bent.

General relativity was Einstein's masterpiece, with consequences so profound that we are still grappling with them today: black holes, gravitational waves, and the whole discipline of cosmology to name just a few. But beyond its implications, the theory was exceptionally beautiful, concise in its assumptions, wide-ranging in its consequences. Einstein himself described the theory as being of 'incomparable beauty.' Spurred by his success with general relativity, he now believed that an even greater, more beautiful so-called unified field theory was out there waiting to be discovered, one that would combine his own theory of gravity with the electromagnetic theory of his hero, James Clerk Maxwell.

Working alone in his study Einstein pursued his vision with ever greater devotion. Over time, his quest led him further and further from the scientific mainstream, and he grew increasingly isolated, working alone on what many of his colleagues suspected was a fool's errand. Einstein himself wrote that he had become 'a lonely old fellow. A kind of patriarchal figure who is known chiefly because he does not wear socks and is displayed on various occasions as an oddity. But in my work I am more fanatical than ever.'*

Einstein was chasing a dream that he would never see realized. He died in 1955, after having made perhaps the greatest contributions to our understanding of nature of any scientist who has ever lived, but having spent (some might say wasted) the last decades of his life on a quixotic quest for unity through beauty.

Einstein was doomed to fail. Not only had he rejected quantum mechanics, a subject that he'd helped to found, he'd also ignored the rapid advances being made in nuclear and particle physics, including the discovery of the strong and weak forces. No unified theory that left them out had any chance of succeeding. What's more, many of the great discoveries in both quantum field theory and general relativity still lay years in the future. The time simply was not ripe.

Fast-forward two decades to the mid-1970s and things had changed dramatically. Charged by their success in unifying the electromagnetic and weak forces (although the actual experimental proof was still a decade away), theoretical physicists began to think big. The next logical step in the unification project was to combine the strong nuclear force with the newly unified electroweak force in what became known as a 'grand unified theory'. A potential candidate was discovered by Sheldon Glashow and Howard Georgi

---

* Einstein was famous for refusing to wear socks, complaining that his big toe invariably made holes in them.

in 1974, based on the symmetry group SU(5),* another one of those local symmetries that we discussed earlier. Amazingly, they found that this relatively simple symmetry not only generated the electromagnetic, weak, and strong forces, it also gave rise to the matter particles – the electron, the neutrino, and the up and down quarks – with exactly the right charges. Along with the fields of the standard model came a bunch of new force fields, but the trouble was that the associated particles were predicted to have absolutely gigantic masses, around 10^{16} GeV, or ten thousand trillion times heavier than the proton, which with today's technology would need a collider so big it would stretch from the Earth to Alpha Centauri.

However, there was a way to test these grand unified theories. The new force fields that they predicted made it possible for protons to decay into antielectrons and a quark–antiquark pair known as a pion. Now the fact that there is still matter in the universe means that this must happen fantastically slowly, with an average lifetime of around a billion billion trillion years. But if you got enough protons together in one place it should be possible to catch a few of them decaying every so often. Happily, there was a fairly straightforward way to do this – dig a giant hole in the ground away from cosmic rays and sources of background radiation, fill it with *a lot* of water, surround it with light detectors, and wait for the occasional flicker in the dark produced by a decaying proton. In 1982–83, two such giant water tanks, one under the Kamioka Mountain (the current site of the larger Super-K experiment) in Japan and the other down an old salt mine on the shores of Lake Erie, began collecting data. However, as the years rolled by, neither of them spied a single proton decay, and before long the simplest grand unified theory discovered by Glashow and Georgi was all but ruled out.

---

* Just as a reminder, the electromagnetic, weak, and strong forces of the standard model appear to arise because of local symmetries in the laws of nature, which are called U(1), SU(2), and SU(3) respectively.

But just as grand unified theories were coming under pressure from proton decay experiments, a sudden fever tore through the theoretical physics world. In the autumn of 1984, a calculation by Michael Green and John Schwarz transformed what had been a relative backwater to *the* hot topic in theoretical physics. Forget grand unified theories; the phrase now on everyone's lips was 'string theory'.

String theory had first been studied at the start of the 1970s as an attempt to understand the strong force that binds quarks together. It ultimately failed in that task, but as time passed it slowly morphed into something far more ambitious, a quantum theory of gravity. During the 1970s, theorists had discovered that string theory contained an object with the precise properties of the graviton, a hypothetical particle that is for gravity what the photon is for electromagnetism. However, string theory's previous failure in describing the strong force led most theorists to treat it with scepticism, that is until the autumn of 1984. Green and Schwarz had managed to show that string theory was free of the mathematical nasties known as anomalies.* A theory with anomalies is a nonstarter, like a sailing ship with a dirty great hole below the waterline, so showing that string theory was anomaly-free suddenly opened up the possibility that it really could be the answer to the long-sought quantum theory of gravity.

The autumn of 1984 was the beginning of what became known in theory folklore as the 'first superstring revolution'. Theoretical physicists piled into the subject, smelling the whiff of the grand synthesis that Einstein had dreamed of. The great promise of string theory was not only as a quantum theory of gravity but as a theory of everything, a single framework that would explain all the features of the subatomic world. What's more, there were hints that string theory might be unique, the kind of perfect final theory that

---

* Not to be confused with the experimental anomalies we were discussing in the previous chapter.

Weinberg would write about in 1992 inspired by string theory's successes over the previous ten years.

Endless books have been written about string theory by people far more expert than I am, so if you want the full lowdown in all its mind-melting complexity then I encourage you to read one of them.* However, for the purposes of our story I'll just outline the key points. At the heart of string theory is the captivating idea that if you zoomed in on a particle like an electron, you would eventually see that it wasn't a particle but a tiny vibrating string. The string is the fundamental building block of everything, with all the different particles in nature corresponding to different ways the string can vibrate. You can think of these like notes on a guitar string: one note gives you an electron, another a quark, another a graviton. String theory turns the subatomic world into a quantum mechanical symphony.

But this beguiling picture comes at a price. First of all, string theory only makes sense if the universe is supersymmetric, which is why it's often described as 'superstring theory'. However, unlike the version of supersymmetry that was introduced to stabilize the Higgs, in string theory the superparticles can have any mass you like, going all the way up to the Planck energy, and so not finding sparticles at the LHC doesn't rule out string theory.

The more serious price for string theory is that it only works if there are at least nine dimensions of space. This might seem like a pretty fatal flaw given that we live in a decidedly 3D world, but again this can be got around by hiding the six extra dimensions way, way down at the Planck length, far beyond the reach of any experiment. You might be starting to notice a theme here. Nevertheless, in its glory days of the late 1980s and early 1990s, there were hopes that at some point in the future string theory would start to make predictions that could be tested in the fire of experiment.

* I can recommend *The Elegant Universe* by Brian Greene.

Over the subsequent decades those hopes have gradually waned. The problem lies in those extra dimensions. To get a string theory that makes statements about the world you first have to hide these extra dimensions through a process known as 'compactification', which more or less corresponds to screwing them up into a tiny, complicated shape, a bit like screwing up a bit of paper into a ball, except this is a piece of hyperpaper with six instead of two dimensions. Anyway, how you screw up the extra dimensions completely changes the kind of universe the theory describes, as their shape determines the different ways that the strings are allowed to vibrate, effectively changing the possible notes you can play on the strings. This in turn gives rise to universes with completely different forces and different particles.

Physicists had hoped that there might be only one unique way of compactifying the extra dimensions, giving rise to one unique theory of the universe. Unfortunately, the number turned out to be much bigger than one. Much, much bigger. Prepare to meet the largest number you are ever likely to encounter except infinity: 10^{500}. That's a one with five hundred zeros after it. I'm not going to write that out in full because my editor will kill me. It is a number so vast that if you wanted to write it down with tally marks – that is, if you wanted to scribble down 10^{500} lines on paper – you couldn't do it. There aren't enough atoms in the universe. Not by a long shot.

This is something of a problem. Imagine you're a string theorist and you want to see if your favourite version of string theory predicts the particles that exist in our universe. You screw up the extra dimensions in your preferred way and calculate the consequences. Oh dear, this universe had eight quarks instead of six. Never mind, though, there are still $10^{500}-1$ other string theories left to choose from. Unfortunately, even if you converted every atom in the universe into string theorists you would never be able to come close to checking all the possible different versions. To date, no one has managed to find a version of string theory that succeeds in describing the particle content of our universe, leading some to refer to it rather unkindly as 'the theory of everything else'.

Weinberg's dream of a final theory seems to have turned into a nightmare; far from being a unique description of our universe, string theory appears to be so flexible that it's impossible to prove wrong. Some still hold out hope that eventually a new principle will be found that shows that there are really only a small number of ways, perhaps even just one way, of screwing up the extra dimensions. However, a more common response is to accept a more limited brief for string theory.

Someone in this camp would argue that it is unreasonable to expect string theory to predict precisely the particles that we happen to find in our universe, just as it would be unreasonable to expect Newton's law of gravity to predict the number of planets in the solar system. Newton could beautifully describe how the planets orbit the Sun, calculate the shape of their orbits and the length of their years, but the exact structure of the solar system – two ice giants, two gas giants, and four rocky inner planets* – is just a historical accident. We know there are hundreds of billions of stars in our galaxy, almost all of them with their own planetary systems, most of which are very different from ours.

This sort of argument works for Newton's law of gravity because we know there are a gigantic number of stars in the universe. However, string theory is making statements about the basic ingredients of the *entire* universe. For this sort of argument to hold, you need there to be multiple universes, potentially around 10^{500} of them, to give ours a decent chance of forming. Accept this, and the fact that we have the fundamental particles that we do is just an accident of history. Some unknown mechanism, presumably at the moment of the big bang, randomly screwed up the extra dimensions in just the right way to give rise to the world we find ourselves in. In most of the other universes the particle content and the laws of

---

* Let's not get into an argument about Pluto.

nature are completely different, and we find ourselves living in the universe we do because the conditions randomly turned out right for our form of life to evolve.

The multiverse is a get-out-of-jail-free card. Not only does it relieve string theory of the duty to explain the universe we live in, it's an all-purpose solution to pretty much any problem you can think of. Why is the Higgs field miraculously tuned such that atoms can exist? The multiverse. How did matter win out over antimatter in the big bang? The multiverse. Why did my mum accept the offer of a vodka and orange from my dad at a British Telecom training course in 1974? You guessed it, the multiverse.

I'm not saying that the multiverse isn't logically possible; if anything, the history of science suggests it could well be true. We used to think that the Earth was the centre of the universe, then we realized, after a bit of arguing, that we were just one planet of several orbiting the Sun. Then our Sun got demoted to just one of a vast number of stars in the Milky Way, and finally the Milky Way turned out to be just one of billions upon billions of galaxies. Philosophically, the idea that our universe isn't unique makes a lot of sense. It's just that we have no way of knowing.

We can't disprove the existence of the multiverse, in much the same way we can't disprove the existence of God. It's true that the effects of other universes *could* show up in the sky if one happened to bump into ours, just as God could unzip the sky one day and give us a cheery wave and/or rain hellfire, depending on your religious preferences (I was brought up Church of England, so in my case he'd offer us tea and a custard cream). But just because we don't see that happening doesn't mean that either God or the multiverse doesn't exist. And God as a hypothesis does an equally good job of explaining why we live in the universe we do.

The multiverse amounts to giving up, throwing our hands in the air, and saying, 'Oh, it's all too hard.' It makes us stop looking for answers, and so, as far as I'm concerned, it isn't worth spending more than a moment thinking about. The multiverse is boring!

Given this rather unpalatable situation, what then is string theory good for? Well, there are many answers to that question. For one, it *is* a quantum theory of gravity, and probably the only one that's been found so far. When you zoom out and look at string theory at long distances it turns into Einstein's general theory of relativity, and when you zoom in, it looks like quantum mechanics, which is an achievement that none of its rivals can yet match. It is more than possible that string theory is the quantum theory of gravity needed to describe the moment of the big bang, and that the standard model has to be bolted on to explain particle physics. While this wouldn't be the idealized theory of everything that many dreamed of, between them these two theories could describe more or less any situation you could care to imagine in the entire history of the universe.*

What is more, it is an immensely rich mathematical structure and a powerful tool. Most people working in string theory today aren't looking for a fundamental theory of everything, or even studying quantum gravity, but are using it to make discoveries in pure mathematics, to better understand quantum field theory, and even to study the physics of solids and quark–gluon plasmas. It is this richness that makes string theory so interesting to the thousands of theoretical physicists and mathematicians who work in the string community. All power to them, I say, and indeed to everyone pursuing other potential approaches to quantum gravity. Compared to experiments, theorists are cheap; they only really need somewhere to sit, an inexhaustible supply of paper and coffee, and a bin.

However, a not altogether unjustified critique of string theory as an approach to a fundamental theory of the universe is that many of its adherents don't worry much about the fact that it has

---

* OK, to avoid being accused of physics imperialism I should say that strictly speaking it could describe any process involving fundamental particles or gravity. If you want to explain anything complicated like biology, economics, or love then physics probably isn't going to be much help.

yet to make any experimentally testable predictions. To be fair, this isn't a problem just for string theory – it's a problem for all quantum theories of gravity. The essence of the issue is this: quantum gravity theories, by definition, describe nature when both quantum and gravitational effects are strong, and this only happens at indescribably extreme energies and densities, those that we believe existed at the moment of the big bang.

The Large Hadron Collider can reach energies of 14,000 GeV. However, to reach the Planck energy, where we would expect to see the effects of quantum gravity, we'd need a collider that could smash particles together with energies close to 10^{19} GeV, a thousand trillion times higher. How big would such a collider need to be if it worked along similar lines as the LHC? About the size of the Milky Way galaxy. Given the current funding climate, I can't see that getting approved anytime soon.

Trying to predict the future is a mug's game, and who's to say that there won't be some incredible breakthrough in accelerator technology that might make reaching the Planck energy possible someday. But I would be happy to bet that it won't happen this century, or even in the next. In fact, I suspect it might never be possible. If that's right, then even if string theory does describe the physics of the moment of the big bang, we will probably never be able to test it in the lab.

However, all is not lost. We may never be able to build the Ultimate Collider, but the universe itself may offer us another way to get tantalizingly close to the Planck scale. For most of the last fifty years we have only been able to look back to a point 380,000 years after the big bang, when the primordial fireball cooled to form a transparent gas, releasing the blaze of light that would fade to become cosmic microwave background. The cosmic microwave background represents a firewall through which ordinary telescopes cannot penetrate. However, as of September 2015 we have a brand-new way to look at the universe, one that may allow us to peer right back to an instant after the beginning.

THE ECHO OF CREATION

Deep in the forests of southern Louisiana, where the warm, humid air makes the loblolly pines grow tall, a revolution in our understanding of the universe is underway. Just outside the small town of Livingston is a telescope like few others on Earth: a huge L made from two 4-kilometre-long concrete tubes that slice their way through the woodland at a right angle, like some giant geometer's tool. It's a strange look for a telescope, but that's because this instrument doesn't study the universe in light. It uses gravitational waves.

To reach LIGO (the Laser Interferometer Gravitational-Wave Observatory), you turn off Highway 190 at Livingston, bump across a poorly maintained level crossing, and wind your way through forestry land, passing the occasional house or trailer, some with broken-down cars rusting in the front yard. Rounding a bend onto the final 500-metre straight leading to the gates of the observatory, a sign instructs you to slow to a 10 mile per hour crawl, hinting at the extreme sensitivity of the instrument ahead.

LIGO burst into the news in February 2016 after announcing the first direct detection of gravitational waves, ripples in the fabric of spacetime predicted by Albert Einstein almost exactly a century earlier. Gravitational waves are a direct consequence of general relativity, which describes spacetime as a dynamic fabric that can be bent, stretched, and squashed by massive bodies like planets or stars. Its elastic nature also allows it to carry waves, ripples that expand and compress space and time as they pass by.

At 5.51 a.m. on the morning of 14 September 2015, just as LIGO was about to begin its first data-collection run following a major upgrade, the Livingston observatory picked up the first-ever signal of a passing gravitational wave. Seven milliseconds later, its twin instrument 3,000 kilometres away in Hanford, Washington, detected the same wobble in the fabric of spacetime as it tore northwards through the Earth at the speed of light. The wave was the echo of a cataclysmic collision between two gargantuan black holes, each around thirty times the mass of the Sun, which had spiralled into each other 1.3 billion years ago in a galaxy far, far away. In the final fraction of a second, the merger produced a disturbance in space-time so violent that it pumped out fifty times more power than the entire visible universe, converting three Suns' worth of mass into pure gravitational energy. However, thanks to its extreme remoteness, by the time this almighty blast reached the Earth 1.3 billion years later, it caused the length of LIGO's two giant arms to flex almost imperceptibly, by just one-thousandth of the width of a proton.

With this first signal, LIGO opened a new window on the universe. For the first time, it became possible to look into a hidden world, to study objects that emit neither electromagnetic radiation, nor neutrinos, nor any other subatomic particles. Colliding black holes and neutron stars, and perhaps things altogether strange and new, are now within reach.

After navigating the security gate, I was met by the head of the Livingston observatory, Joe Giaime, in front of the main LIGO building, a large metal-clad warehouse painted in two horizontal bands of blue and white to help it blend into its surroundings. Joe has been working on LIGO his entire career, starting out as a technician at MIT back in 1986. His doctoral supervisor was Rai Weiss, one of the founders of LIGO who would go on to share the 2017 Nobel Prize in Physics with Kip Thorne and Barry Barish for the discovery of gravitational waves.

Back in those early days, Joe didn't realize quite how special the project he was embarking on was. He started a year before the name 'LIGO' was even coined and helped put together a joint MIT–Caltech proposal that was submitted to the National Science Foundation in 1989. Just six years later, they had funding and were breaking ground at the twin sites in Louisiana and Washington, lightning fast for such a big science project.

Although he formally works in astrophysics, Joe spent thirty years without making a single observation of the heavens. He describes himself as an instrumentalist by inclination. 'I cut my teeth building and designing things,' he said. Until 2015, his entire career was devoted to getting LIGO to the point where it would finally be able to start studying the universe.

Joe led me from the main building on a short walk to a bridge spanning one of LIGO's giant arms. From there we could look in a dead-straight line through the forest to a point where the concrete tube met its end station, 4 kilometres in the distance. To our left, a second arm emerged from the main LIGO building and plunged through the forest at a right angle.

LIGO works by detecting tiny changes in the lengths of the two arms as a gravitational wave causes space to expand and compress as it passes by. Inside the main building, a laser is split into two beams that are fired down the two perpendicular arms, which then bounce off mirrors at the end stations and back down the same tubes to recombine in the main building. Generally, a gravitational wave will change the length of one arm more than another, so that when the lasers recombine, the positions of their peaks and troughs are ever so slightly out of whack, producing what is known as an interference pattern.

At least, that's the idea. But the effect of a passing gravitational wave is so minute that it can be easily overwhelmed by vibrations from all manner of things here on Earth. The forest surrounding the Livingston observatory is owned by an international wood and paper company – the hot, humid Louisianan climate means that trees grow unusually fast here – and the felling of trees is an occasional source of background noise (to say nothing of British science writers and their noisy hire cars). Nonetheless, LIGO manages to live in what Joe described as 'uncomfortable harmony' with the local logging industry.

It's not just falling trees that LIGO has to contend with. The instrument is sensitive to impossibly tiny changes in the length of the arms, all the way down to 10^{-19} metres, one ten-thousandth of the width of a proton, or 'the private space enjoyed by two quarks', as Joe put it. However, there is a long list of sources of vibration that could shake the optics by far larger amounts, everything from footsteps in a nearby corridor to waves crashing on the continental shelf in the Gulf of Mexico. LIGO copes with all of these through an ingenious system of seismic isolation, including sets of quadruple pendulums that keep the mirrors as still as possible.

LIGO first achieved sensitivity in 2005, shortly after Hurricane Katrina devastated the nearby city of New Orleans and the surrounding area. Even in those early days there were hopes that LIGO might see a signal, but it would take a further ten years of laborious upgrade work finally to reach the level of precision that allowed them to catch their first wave.

Back in the main building, Joe showed me into the control room, with banks of desks and computer monitors facing larger screens on the front wall. Just as we entered the room there was a commotion among the staff, with some standing up from their desks to study the screens in more detail. 'We've lost lock,' said Joe. Just at that moment, a seismic wave from a 7.1 magnitude earthquake near Indonesia's Maluku Islands had hit LIGO, which even at a distance of 15,000 kilometres was enough to shake the optics out of alignment. 'We won't be able to do anything now for several hours while that dies down,' Joe told me, 'while those waves go round and round the Earth.' Standing in the control room I couldn't help but marvel that any of this worked at all. To be able to measure length changes ten thousand times smaller than a proton, while contending with tremors from all and sundry, including earthquakes on the other side of the planet, seems nothing short of miraculous.

Nevertheless, it does work, and beautifully. Despite operating for just a few short years, LIGO is already starting to change our understanding of the universe. Perhaps the most significant event so far was detected on 17 August 2017, nearly two years after it landed its first gravitational wave. This time, both LIGO observatories, along with their European counterpart, Virgo, in northern

Italy, picked up a signal from a collision between two neutron stars, the ultradense husks left over from violent supernova explosions. As soon as the gravitational wave was detected, LIGO and Virgo sent out an alert to telescopes all over the world, which started to scan the sky feverishly for an accompanying electromagnetic glow. Unlike black holes, a collision between two neutron stars is expected to produce a powerful burst of electromagnetic radiation – which was indeed spotted eleven hours later, emanating from a galaxy 140 million light-years from Earth.

Not only was this the first time a gravitational and electromagnetic signal had been detected from the same collision, it also made astrophysicists re-evaluate their ideas about where the chemical elements come from. As we saw, for a long time it was thought that the heavy elements beyond iron were made when giant stars went supernova. However, there had been a growing suspicion that neutron star mergers might in fact be their main source. Sure enough, spectroscopic studies of the light from the 2017 collision revealed telltale signs of the production of precious metals, including gold and platinum, suggesting that a large fraction of the metal in a piece of jewellery came from exactly such a collision.

Back in Joe's office with coffees in hand, we talked through LIGO's plans for the next few years. 'There's a scaling law that's wonderful and horrible,' he explained. Every time you double the instrument's sensitivity, you can see twice as far into space, but because the volume of space you can scan increases with the cube of the range of the instrument, the number of events you can detect goes up by a factor of eight. This creates a temptation to always be making improvements instead of collecting data. 'Everybody has this itch to make tiny little changes. You can convince yourself that the payback is so enormous that essentially you should never be running.'

In practice, they take a more pragmatic approach and spend half the time recording data and half improving the instrument, with an aim to double LIGO's sensitivity by 2024. This will bring a huge unexplored region of the universe into view. But in the longer term there are even grander plans in the works.

By proving that gravitational waves really exist, LIGO effectively invented an entirely new type of astronomy. A number of large projects are now being planned that could have truly revolutionary impacts on our understanding of the universe and its history. Europe is currently drawing up a proposal for the Einstein Telescope, a huge underground triangular observatory with three 10-kilo-metre-long arms, while in the United States a supersized version of LIGO with 40-kilometre arms known as the Cosmic Explorer is being studied. But perhaps most ambitious of all is LISA (Laser Interferometer Space Antenna), three spacecraft flying around the Sun in an equilateral-triangle formation, firing laser beams back and forth between them, effectively creating an observatory with arms that are 2.5 million kilometres long. Having spent years in the doldrums, the LISA project has been reinvigorated by LIGO's discovery of gravitational waves, with the European Space Agency planning to launch the mission sometime in the 2030s.

Joe told me that these telescopes will be so sensitive that they will be able to see every black hole collision in the observable universe and look right back to the time when the first black holes formed from dying stars. One extremely exciting possibility is that they might discover a population of primordial black holes that didn't form from collapsing stars but during the big bang itself. In the first second, when the universe was extremely hot and dense, it's possible that fluctuations in the quantum fields could have created regions that were so dense that they collapsed into black holes, which in principle could have survived to this day. If the Einstein Telescope or the Cosmic Explorer saw black hole mergers taking place before the first stars formed, that would be an unmis-takable smoking gun of their existence. Another possibility is that it could find black holes that weigh less than the Sun, a weight that marks them as too light to have formed from a collapsing star. Discovering primordial black holes would be huge, not only telling us about the conditions in the very first moments of the big bang but potentially also providing an explanation for some of what makes up dark matter.

However, perhaps the greatest prize of all would be seeing back into the fireball of the big bang directly. Until around 380,000 years after time zero, the entire universe was filled with a searing-hot plasma of subatomic particles. This fireball was opaque to light – any photons flying about before this time would have been endlessly pinballing off protons and electrons – which means that we can't see back any further than this with ordinary telescopes. Gravitational waves, on the other hand, don't get absorbed by matter and so would have been able to zip unimpeded through the universe from its very earliest moments.

For gravitational waves from the early universe to still be detectable today they would have to have been produced by unimaginably violent processes. One possibility we've already met: the collision between expanding Higgsy bubbles around a trillionth of a second after the big bang. This was the idea that matter may have won out over antimatter thanks to the Higgs field turning on unevenly through the universe, forming bubbles in the hot plasma. As these bubbles grew, they would have smacked into one another with incredible force, sending powerful ripples through the fabric of spacetime, the faint echoes of which could be picked up by one of the next generation gravitational wave observatories. If future astronomers were able to detect such a signal it would tell us directly about the physics going on in the first trillionth of a second and potentially help us unravel the mystery of where the matter in our apple pie ultimately came from.

But maybe, just maybe, we'll be able to see back even further. We said earlier that it's almost impossible to imagine being able to build a particle collider powerful enough to study quantum gravity. But if you go back far enough perhaps the entire universe once acted as the ultimate collider. This time is believed to have occurred around a trillionth of a trillionth of a trillionth of a second after time zero, when the universe underwent a short period of extremely rapid expansion known as 'inflation'.

Exactly how inflation happened, or indeed if it happened at all, is still uncertain, but it is thought that in an incredibly short time – a mere ten-billionths of a trillionth of a trillionth of a second – the universe ballooned to at least 10 trillion trillion times its previous size. To put that in some kind of perspective, if the period at the end of this sentence grew by the same factor it would end up a hundred times larger than the Milky Way.

Inflation explains a number of peculiar features of our universe, in particular why it looks remarkably uniform whichever direction you look in. This is really surprising because without inflation, two opposite regions of the sky should never have had been in contact long enough to have reached the same temperature and density. Inflation solves this problem by saying that even two points on opposite sides of the observable universe were once part of the same tiny patch of space. Perhaps even more remarkably, inflation also says that all the large-scale structure in the universe – in other words the way galaxies are distributed through space – ultimately came from tiny quantum fluctuations that occurred at distances far smaller than an atom and that then got blown up by inflation to absolutely enormous scales. These quantum fluctuations resulted in some areas of the universe being slightly denser than others, and these overdense regions eventually went on to collapse under gravity to form everything we see when we look up at the night sky. In other words, the billions of billions of galaxies in the observable universe were ultimately seeded by tiny wobbles down at the quantum level in the first instant of cosmic time.

Inflation is a generally accepted part of the cosmological story, and while many of its predictions have been confirmed there is still no unequivocal evidence that it actually happened. There's no way to look back directly to a trillionth of a trillionth of a trillionth of a second after the big bang with ordinary telescopes, but the existence of gravitational waves may now make it possible. If inflation really did happen, it would have roiled spacetime, creating wild waves in the fabric of reality that should still be echoing through the universe. Today they would be stretched to incredibly long wavelengths and be impossibly faint, but nonetheless there is a chance that future planned observatories may be able to catch these whispers from the birth of the universe.

The trouble is that inflation isn't just one theory – there are a whole bunch of different ways that inflation could have happened, each involving different numbers of quantum fields and different energy scales – and only some versions of inflation would produce powerful enough gravitational waves to be picked up directly. In the very simplest models, the waves would be too weak for even the giant LISA space observatory to detect. In that case, the one chance we might have to hear the echoes of inflation is to look for their effect on the oldest light in the universe, the cosmic microwave background.

Theorists have calculated that gravitational waves produced by inflation would have left twisting patterns, known as 'B-modes', in the cosmic microwave background. These are fiendishly difficult to detect, suffering from the double whammy of being both extremely weak and easily confused with more mundane backgrounds, like dust in our own galaxy. In fact, in 2014, the BICEP2 telescope at the South Pole caused a worldwide sensation when the team announced that they'd seen evidence for twists in the cosmic microwave background caused by gravitational waves from inflation. Breathless talk of a new age in our understanding of the cosmos and the imminent award of Nobel Prizes abounded, but as time passed the BICEP2 team were forced into a humiliating climb-down. It gradually became clear that they hadn't properly accounted for the effect of galactic dust, and after reanalysing their results the claimed signal melted into the background.

Despite this false dawn, there are good prospects that the imprints of gravitational waves in the cosmic microwave background will finally be detected in the coming years. A range of ambitious new telescopes, at the South Pole, high in the Atacama Desert, and orbiting the Earth, will produce maps of the cosmic microwave background so exquisitely detailed that they should be sensitive enough finally to spy the effects of primordial gravitational waves, if indeed they exist. If that were to happen, it could be our best chance to get actual data on the very first moments of the universe's existence and the very highest energies imaginable.

Just before I left Joe's office to head back to my hire car, he had a treat for me. 'Take a listen to this,' he said, as he fiddled with his computer. After a few moments of quiet, I was startled by a loud, unearthly rumble coming from a subwoofer hidden at the back of the room. 'This is the sound of the first gravitational wave.' As I listened, rising just above the deep rumbling came an abrupt thump, the sound of two black holes crashing headlong into each other, 1.3 billion years ago.

It's easy to become inured to the achievements of modern science, but sitting in Joe's office next to one of the most sensitive instruments ever built, listening to the echo of an event so remote in time and space and yet so vast and violent that it defies all description or imagination, I couldn't help but feel a rising sense of optimism. Science is exploration, whether it's done in the laboratory, the abstract world of mathematical theory, or by studying the signals from the universe itself. And as we explore we seem to endlessly stumble upon new phenomena and new mysteries that lead us farther and farther from where we started. Does this journey go on forever, or will we one day reach its end? That, perhaps, is the biggest question of all.

CHAPTER 14

The End?

It is the year 843 million CE. After a hundred thousand millennia of construction, the Galactic Organization for Particle Physics (GOPP) calls a press conference to inaugurate its last and greatest scientific project. Hanging in the void of space, glistening like a circlet of silver around the centre of the Milky Way, is the largest, most powerful, and most expensive machine ever built in the history of the observable universe – the Impossibly Large Hadron Collider. Three thousand light-years in circumference, the ILHC is the work of a pangalactic collaboration of more than eight hundred thousand intelligent species who have put aside their differences in an attempt to discover the fundamental nature of reality. Today is the day the whole galaxy has been waiting for, when at long, long last their gleaming new machine will collide particles with enough energy to probe the effects of quantum gravity. A complete understanding of the fundamental laws of nature is finally within reach.

It has been a long and bumpy road; centuries of grant proposals and funding applications, endless quibbling over which star systems would be awarded the key magnet contracts, not to mention more than one thousand court cases filed across the galaxy over fears that the collider will trigger the end of the universe. Even that very morning, the announcement had to be delayed after the French delegation demanded that the press release be made available in their own ancient language as well as in the far more widely spoken Galactic Creole.

Nonetheless, the moment has finally arrived. It has taken just over a million years to accelerate the protons to the required energy of 10^{19} GeV, and we are now just seconds from the first collisions. The director general of GOPP calls the room to attention with a wave of one of her twelve purple tentacles. 'Ladies, gentlemen, incorporeal energy beings, and sentient fungi, the moment you've all been waiting for. I give you, the Planck scale!' Immediately, the screens around the large conference hall light up with fireworks of particles emanating from a point deep in the heart of the planet-sized detector. 'Professor Splurg, the results if you please.'

A glowing orb of ethereal light approaches the lectern and hands a ticker-tape printout of the data analysis to the eager director general. 'Ummm . . . well, this is interesting,' stutters the DG in a weak attempt to mask her alarm. 'It seems we've made a black hole. Not to worry, though; perhaps if we put in a bit more energy . . . Professor Splurg, more power!'

With a straining of electromagnets, the ILHC pushes its protons beyond the Planck energy, to an incredible 10^{21} GeV. Yet more collisions flash across the screens surrounding the bemused journalists. 'Ah, right, I see . . . ,' stammers the DG. 'Ladies, gentlemen, et cetera, my apologies . . . we will have to bring today's proceedings to an end for now. I need some time to consult with my colleagues.'

This little bit of sci-fi silliness is an attempt to make a serious point: we may never be able to find out how the universe began. Even if we could build the ultimate collider to probe what happens down at the Planck length, we would end up squeezing so much energy into such a small space that we'd collapse our two particles into a black hole. The interior of a black hole is surrounded by a barrier known as the 'event horizon', from which nothing, not even light, can escape. As a result, what was happening down at the

Planck length would be hidden behind the event horizon. Collide particles at even higher energy and the problem gets even worse – you just create a bigger black hole.

This point was made to me even more starkly by David Tong, a professor of theoretical physics and one of the stars of Cambridge's Department of Applied Mathematics and Theoretical Physics, as we sat in his office on an overcast spring afternoon. Aside from being one of the world's leading experts on quantum field theory, David is a captivating speaker, his excitement and curiosity shining through with every word, which coupled with his outward youth and thick-rimmed glasses puts you in mind of a real-life version of David Tennant's Doctor Who.

David started out working on string theory but eventually got put off the subject by the fact that it's almost impossible to test. 'We've got to be extremely lucky to find any evidence for quantum gravity in an experiment,' he told me, 'and it's not going to happen in my lifetime, which makes it a little bit uninteresting.'

Then, with a mischievous sparkle in his eye, he went further: 'If you really want a conspiracy theory, why is quantum gravity uninteresting? There are three things in nature, fundamental laws of physics, that suggest that quantum gravity is fundamentally something we cannot probe, or at least, nature does a very good job of hiding it.'

The first comes from the work of Kenneth Wilson, one of the greatest and possibly most underappreciated theoretical physicists of the twentieth century. Wilson is famous among particle physicists for his work on the renormalization group, a mathematical object that tells you how a system looks when you zoom in or out. Wilson's insight was that in some sense *it doesn't matter* what's going on deep down if you want to understand how a system behaves at larger distances. Or as David put it, 'Newton didn't need to know about quarks to figure out how planets work.'

In other words, it doesn't matter what the fundamental ingredients of the universe are down at the Planck scale; they're very unlikely to leave any traces on the behaviour of much bigger things that we can actually measure in the lab, like atoms or particles. Given that the whole subject of this book is an attempt to understand an apple pie by zooming down and down, that gave me some serious pause for thought.

'Number two, inflation in the early universe, what does inflation do? It just dilutes everything, makes sure any hint of what happens at the big bang is pushed way outside our cosmological horizon and just can never be seen.' Although inflation might conceivably create gravitational waves that we could detect evidence for, these would likely only ever allow us to see back to 10^{-36} seconds after the big bang, when inflation roared into action. The moment of the big bang itself, at time zero, gets dragged far beyond our view, over the horizon and out of sight, by the rapid expansion of space. Inflation hides the 'b' of the big bang from us.

'Number three is cosmic censorship. Where can you hope to really learn about quantum gravity? Well, it's at singularities at the centres of black holes, but they're always hidden behind the event horizons! Gravity is weird; usually if you want to probe smaller distances, you build a bigger and bigger collider. But suppose you build a collider a hundred times the Planck scale, 10^{21} GeV, we know what's going to happen, you collide them together, you form a big black hole.

'So, you wanna build an apple pie from scratch?' David asked. 'Well, the scratch is hidden.'

Way back in the summer of 2011 when I was still a PhD student, I got to attend my first big international conference in the pretty midwestern city of Madison, Wisconsin. The LHC was only a year into its first run, the discovery of the Higgs was still another year in the future, and the presentations mostly consisted of preliminary results – *No sign of supersymmetry yet, folks, but we'll surely catch it soon* – and speculative theoretical proposals. I admit to finding the long plenary sessions tedious at times, that is until I was shaken awake by the arrival onstage of the headline speaker of the conference, Nima Arkani-Hamed.*

Nima speaks with a passion that makes you sit up and pay attention. Dressed head to toe in black, with a mane of black hair swept backwards, he paced the stage like a lion eager to burst out of its cage, as a vision for the future of fundamental physics spilled out of him like a torrent. Barely pausing for breath, he spoke way over his allotted time and deep into the lunch break, but nobody seemed to mind. You couldn't help but get swept along on the wave.

Based at the Institute for Advanced Study in Princeton, where Einstein spent his twilight years and which today is the Holy See of fundamental theoretical physics, Nima Arkani-Hamed is one of the world's most influential physicists. He's famous both for his many contributions to particle theory and for his charisma as a communicator, which puts him in high demand. So I was pretty pleased when I managed to get him on the phone for a conversation about apple pie, as he took a train from Princeton to New York. The first thing he said was characteristically startling: 'I just want to mention one very cool thing that's going on at the moment, one of those quiet intellectual revolutions that frames the subject for the next fifty years or more: we know the reductionist paradigm is false.'

'Oh,' I replied.

Reductionism is the idea that you can explain the world by breaking it down into its basic ingredients. It's the philosophy that underpins particle physics. The entire story that I have been telling over the last fourteen chapters is the story of reductionism. It's an approach to the world that has served us incredibly well for almost half a millennium. So the idea that it's false is, to say the least, a big fucking deal.

The first challenge to reductionism comes from the expectation

---

* The same guy who bet a year's salary that the Higgs would be found.

that if you collide two particles together with enough energy to probe the Planck scale you make a black hole, and if you try to keep going after that, go to even higher energies, you make even larger black holes. 'This is one of the absolutely deepest things that we know about quantum mechanics and gravity,' Nima told me, 'that in a very real sense, higher energies start turning into longer distances again, which is utterly mysterious from a reductionist point of view.'

Now arguably reductionism breaking down at the Planck scale might not worry us too much. After all, the Planck scale is far, far beyond our experimental reach at the moment. But what is surprising, shocking even, is that reductionism might fail us long before we get there. We could be watching it unravel right now at the LHC.

As we've seen, one of the biggest outstanding problems in fundamental physics is the fact that the Higgs field has a uniform value of 246 GeV everywhere, a Goldilocks number that gives particles nice sensible masses, allowing atoms – and thus our universe – to exist. Apart from the untestable multiverse, all the solutions to this problem imply that we should see new stuff as we zoom down to smaller and smaller distances, and therefore higher and higher energies. This new stuff could be superparticles, extra dimensions of space, or perhaps smaller building blocks inside the Higgs itself. However, so far at least, as the LHC has zoomed in on the vacuum, all we've seen is . . . the Higgs.

To borrow Ben Allanach's metaphor, it's akin to walking into a room and seeing a pencil standing upright on its tip. Confronted with such a weird situation, a reductionist would assume there must be something at shorter distances that we can't see that's keeping the pencil upright. Maybe there's an ultrathin wire suspending the pencil from the ceiling, or a tiny invisible clamp that you can only see with a microscope. Not finding new stuff to stabilize the Higgs suggests that this approach is wrong; we can't explain some features of the world by zooming in further.

'The real gauntlet that's being thrown down here by the results of the LHC is a challenge to the reductionist paradigm,' Nima said, 'but much closer to home, in a place that we didn't expect.'

This is what's really at stake now in fundamental physics – the very idea that we can continually learn more about the world by looking deeper and deeper. If reductionism turns out to break down when we try to understand the Higgs, it would be an enormous shake to the foundations of physics. For Nima, studying the Higgs 'to death' is the most important task facing particle physics for the next half-century.

'We have never seen anything like the Higgs before. This is not hyped, it's not like we're making a big deal about the latest particle. The Higgs is the first elementary particle of spin 0 we've ever seen, it's the simplest elementary particle we've ever seen, it doesn't have any charge, the only property that it has is mass, and the very fact that it's so simple is what makes it really theoretically perplexing.'

Since the Higgs's discovery in 2012, ATLAS and CMS have been gradually improving our understanding of it, confirming that it really is spin 0 and measuring how it decays to the other particles. In the mid-2020s, the LHC will get a major upgrade to increase its collision rate, allowing physicists to zoom in on the Higgs even closer. However, by the time the LHC finally powers down around 2035, we will still only have a fairly fuzzy picture. To settle this issue once and for all may require an even more powerful microscope.

Nima has spent much of the last few years criss-crossing the globe, making the case for a successor to the LHC. Two potential projects have emerged as the leading contenders, one based at CERN and the other just outside Beijing. These machines would be true behemoths, around 100 kilometres in circumference, more than three times longer than the LHC and ultimately capable of accelerating particles to energies seven times higher. The CERN project is known as the Future Circular Collider (though presumably it'll get a rebrand if it gets built) and comes in two stages. First a 100-kilometre tunnel would be bored through the Geneva-basin area – the largest that the geology can accommodate – running from the foothills of the Alps, under Lake Geneva, past the current CERN site, and all the way to the Jura Mountains. Into this enormous ring would go an electron–positron collider, designed to make huge numbers of Higgs bosons and study their properties in exquisite detail. Then comes the real monster, a proton–proton collider like the LHC, which would be able to reach collision energies of 100 TeV (trillion electron volts).

Between them, these vast machines would offer a wealth of opportunities for new discoveries in the quantum world. To name just a couple, the proton collider would be so powerful that it would be able to rule out more or less completely the most popular form of dark matter* and be able to recreate the conditions that may have caused matter to form in the early universe. However, as far as Nima is concerned, studying the Higgs is far and away the most important target of these machines and more than enough justification (scientifically at least) on its own.

Of course, 100-kilometre particle colliders do not come cheap. The full Future Circular Collider project would cost an eye-watering €26 billion. But to put that in perspective, it's significantly less than it cost to put humans on the Moon (around $152 billion, or €124 billion, in today's money), and it would be spent over a period of around seventy years, with the proton machine completing its mission close to the start of the twenty-second century. And of course, such a project could only be realized by a collective global effort of dozens of countries pooling their resources over many decades. When spread out like this, it's argued that the Future Circular Collider could fit within CERN's existing annual budget, which costs each citizen of the United Kingdom around £2.30 per

---

* Technically what are known as 'weakly interacting massive particles' (WIMPs), produced in the fireball of the big bang.

year, or as the physicist Andrew Steele pointed out, about the cost of a packet of peanuts.

Even so, we are still talking huge sums of money at a time when the world is facing an unprecedented economic and health crisis. Making the case to spend billions on what might be regarded as a big toy for physicists could easily come across as arrogant or at the very least extraordinarily tone-deaf. Indeed, there is ample warning from history about the perils of such megaprojects. Under the desert near Fort Worth, Texas, are more than 20 kilometres of abandoned tunnels, dug for the Superconducting Super Collider, a 90-kilometre machine that would have reached three times the energy of the LHC. Driven partly by concerns over its ballooning budget, the US Congress cancelled the project in 1993 after more than $2 billion had already been spent, a blow from which American high-energy physics has never really recovered.

I haven't gone into the whole 'why particle physics is good for you' argument so far, because that's not the story I'm telling. However, if there is to be another generation of colliders, physicists now need to engage fully with making the wider case to the general public, beyond the excitement of finding out more about the world we live in. A persuasive argument can be made. First of all, these large high-tech projects invariably generate spin-off technologies that find broad applications, perhaps the best example of which is the World Wide Web, developed by Tim Berners-Lee at CERN as a way to share information between physicists and then given away to the world for nothing. The Web alone has paid for CERN many, many times over. Likewise, superconducting magnets developed for accelerators have found their way into hospitals in the form of MRI machines. Another argument is the inspiration provided by projects like the LHC, with a majority of physics students citing the excitement of particle physics and astronomy as the reasons they went into the subject, most of whom will eventually end up applying their skills in other areas of the economy. And finally, we shouldn't ignore the possibility of one day making use of the fundamental knowledge itself. When J. J. Thomson discovered the electron in 1897, it was regarded as a mere scientist's plaything, and yet today almost all our technology relies on a deep understanding of electrons. Such applications often come long after the fundamental knowledge and are inherently unpredictable, but when they do come, they can be transformative. As ATLAS physicist Jon Butterworth mused, who's to say we won't one day be able to fly across the universe on an interstellar Higgs drive?

That said, just from a scientific perspective it's more than reasonable to ask whether two giant colliders are a good use of €26 billion. Could that money be better spent on other, smaller projects? Maybe, but this kind of assumes that if we weren't going to spend €26 billion on colliders, we'd get to spend it on other areas of fundamental research. Unfortunately, that isn't really how the world works. CERN has been uniquely successful in persuading governments to commit resources to fundamental research for decades now, partly thanks to its many successes but also because of the international prestige that comes from being part of a world-leading scientific organization. The idea that if CERN shut up shop its budget would get redistributed to other areas of research is seriously naive. In the end it really comes down to whether you think trying to answer these big questions is worth the price tag when balanced against the other potential benefits.

A scientific argument that's been made against these machines is that there is no reason to expect them to find any new particles. The people behind the LHC promised that we'd discover supersymmetry and dark matter, and yet, so far at least, they haven't delivered. Or so the argument goes. When I put this to Nima, I could sense his blood rising; I could only imagine the reactions of his fellow passengers on the train bound for New York as he grew increasingly animated.

'That, I think, is an especially dumb argument. It comes from people who got into particle physics because they wanted to see some new bumps in a plot and go to Stockholm or something. And that's what they thought particle physics was about. And they say, look, it's even built into the name of the goddam subject! This is just the honest truth. For me that was not the attraction of the subject *at all*, it made it feel a bit like chemistry, and I sucked very badly at chemistry. And you know, all these particles, all these funny names, were actually a barrier to me that I had to overcome. But of course, what got me into it is you get this most amazing view of the deep workings of the laws of nature. That's what it's really about!

'There's this cognitive dissonance,' he continued. 'People like me will go around saying this is the most amazing period in physics for a hundred years and then there are other people going, "Oh my god, it's so depressing, we've only seen the Higgs and nothing else." And it can be confusing to hear these two different things. Like, am I on drugs? Or are they on drugs? My attitude is that this is a great time, we know that there's some 90-degree turn in the trajectory of the subject. I think it's a 90-degree turn into the most profound place we've been in a hundred years; other people might think it's a 90-degree turn into darkness and death. And the people who think that should do something else with their lives.'

Decades from now, if a future collider discovered supersymmetry or perhaps found that the Higgs is really made of smaller things, the long march of reductionism would go on. Once again, we would have understood more about the world by looking deeper. However, bizarrely perhaps, the most exciting outcome of all would be if these giant machines found nothing at all. No supersymmetry, no extra dimensions, just the plain old fundamental Higgs boson. Reductionism would have failed, forcing a radical rethink of our entire approach to understanding the world we live in. You might wonder, why not just ask a theorist to assume that we find nothing new and then figure out how to deal with that? The problem is, you can't start a revolution without knowing for damn sure that the old regime needs overthrowing. Or as Nima put it as he jumped into a cab at Penn Station, 'You need experiment to turn your entire fucking world upside down.'

Over an unusually hot, sunny weekend as the summer of 2019 gave way to autumn, CERN threw open its doors to the general public. In just two days, more than 75,000 people thronged to see the giant experiments of the Large Hadron Collider, sometimes queuing for several hours in the baking sun for a chance to take a trip underground. Dressed, rather fetchingly I thought, in a bright blue high-viz jacket, luminous orange T-shirt, and the obligatory hard hat, I took group after group down the 100-metre lift shaft to stand beneath the towering LHCb experiment. While it probably wasn't obvious to the visitors peering into the confusing jumble of multicoloured metalwork, most of the detector was missing.

When the LHC switched off for its second long shutdown at the end of 2018, my collaborators on LHCb began a two-year project to more or less completely replace the experiment. When the LHC restarts again (we hope in 2022), the upgraded LHCb experiment will be able to record data at forty times its previous rate, allowing us to home in on ever-rarer processes. As I write this, the anomalies in how beauty quarks decay that have caused so much excitement over the past few years are still there, and early in 2020 some of my colleagues released a result that seems to show that they were strengthening. It's still too early to say which way things will go, whether the anomalies will fade away or whether we will soon have convincing evidence of new quantum fields beyond the standard model. But the LHCb upgrade will give us the crucial data we need. There is a chance, albeit an uncertain one, that a major step forward in our understanding of matter is on the horizon.

We're living in a golden age of physics and cosmology, where experiments and observatories that would have been almost unimaginable just a few decades ago are teaching us more and more about the universe we live in. As I write this, the team at Borexino under the Gran Sasso mountains have just announced that against all odds they've snared their final great prize: neutrinos produced by the carbon–nitrogen–oxygen cycle that cooks protons into helium at the centre of the Sun, filling in yet another part of the story of matter's origins.

The future is bright. Over the coming decades new gravitational wave observatories, telescopes on Earth and in space, dark matter detectors deep underground, precision laboratory experiments, and gigantic neutrino observatories will come online. No one can say what they'll find – experimental physics is exploration – but there will surely be surprises. In my own field, the LHC has a decade and a half still to run, and thousands of physicists will continue to determinedly scour the trillions upon trillions of collisions in the hope of finding clues that may lead us to the next layer of reality.

Reading popular science books and watching documentaries as a kid growing up in the 1990s, I got the sense that physics was hurtling towards a dramatic climax. That after a century of revolutionary discoveries and ever-more-unified theories, physicists were on the brink of finding the ultimate theory of the universe. Huge progress has been made since then, and yet Einstein's dream has if anything drifted further out of reach.

Perhaps it was hubris. The 1970s and 1980s were a time of miracles: forces were unified, predictions spectacularly verified, and beautiful new mathematical structures uncovered. Perhaps all that success made people think that we were ready to jump from the standard model all the way to a theory of everything. In any case, it hasn't turned out that way. Today we can explore physics at incredible energies of around 10,000 GeV in the lab, but the Planck scale is a thousand trillion times higher than that. To think that we could leap from the solid ground of experimental evidence fifteen orders of magnitude into an unexplored world of quantum gravity in a single bound seems to have been premature, to say the least.

Will we *ever* learn how to make our apple pie from scratch? Quantum mechanics and gravity seem to be telling us that the moment the universe began – when gravity, space, time, and the quantum fields of nature were all united – might be inherently unknowable. But that's no reason to get downhearted – the opposite, in fact. We have come a very long way in our understanding of the basic ingredients of matter and their cosmic origins. But we have a very, very long way to go yet before we reach the Planck scale. Putting aside dreams of a final theory, there are many big mysteries left to solve that are closer at hand. What the hell is dark matter? How did matter survive annihilation during the big bang? Can we explain the weirdness of the Higgs field? Science thrives on such mysteries, and these are all questions that we have a chance of answering in the coming years.

When I spoke to Nima, he said that the biggest bottleneck to building the next generation of colliders isn't the money, or persuading politicians or the public, nor is it the formidable engineering challenges. The biggest bottleneck is whether there is a generation of young people who are willing to devote their lives to understanding the Higgs boson. Among the many groups of eager visitors whom my colleagues and I showed around LHCb that weekend were dozens and dozens of teenagers who had given up their free time to squeeze into a lift with some berk in a hard hat and spend an hour or more gawping at bits of scientific equipment. I came away from the weekend feeling pretty optimistic for the future. Someday, years from now, perhaps one of those young people will find themselves sitting nervously in an early morning run meeting as the Future Circular Collider prepares to fire up for the first time.

If so, they will be part of a story that stretches back centuries, the story of how we have gradually understood the building blocks of matter and their origins. It's a story that I fell in love with as a curious teenager and have never stopped being thrilled by. Beyond the incredible discoveries – and who could help but be seduced by the idea that we are made of stuff forged inside stars and in the heat of the big bang – is the fact that thousands of people, across time and cultures, working in different fields, all with their own dreams, strengths, weaknesses, and egos, have slowly built on the achievements of those who came before and led us to an ever-deeper understanding of the world we all share. Most of them never knew one another and were struggling with their own small part of the puzzle, and yet somehow they wove one tapestry, one story, which as far as I'm concerned, at least, is the greatest ever told.

Even an object as mundane as an apple pie is deeply rooted in this cosmic drama, and to truly understand it is to understand the universe and our own small part in it. There may be good reasons to think that we will never be able to discover its ultimate origins, but then again, nature has shown an almost infinite capacity to surprise us. As we go on exploring, gazing farther into space and probing deeper into the smallest elements of matter, who can say what new wonders we will find? We have come a long way, but the story isn't done yet. It is still being written. If we carry on exploring, then perhaps one day we'll finally find the recipe for our universe.

How to Make an
Apple Pie from Scratch

Serves eight. Preparation time 13.8 billion years.

Ingredients:

A smidge of spacetime
Six quark fields, six lepton fields
U(1) × SU(2) × SU(3) local symmetries
One Higgs field
Supersymmetry or extra dimensions of space
   (depending on taste)
Dark matter (not available in shops)
Probably some other stuff

Instructions:

First, invent the universe.

Inflate your initial smidge of spacetime for approximately 10^{-32}
seconds until your universe has grown to around ten trillion
trillion times its original size. Take care not to allow inflation to
continue for too long or you will just end up with a howling
void, spoiling the dish.

After inflating, you should find that your universe's temperature rises dramatically, creating large numbers of particles and antiparticles. Meanwhile, your U(1), SU(2), and SU(3) local symmetries should automatically produce the electroweak and strong force fields. Allow to continue to expand and cool at a gentler rate for another trillionth of a second.

At this point, begin to switch on the Higgs field, aiming for a value of around 246 GeV. I recommend using supersymmetry or extra dimensions to ensure the field remains stable, otherwise you will find it almost impossible to cook atoms later on. However, if you prefer, you can simply repeat the above instructions approximately a million trillion trillion times until you randomly get it right.

To make matter, try to ensure that the Higgs field switches on unevenly, forming expanding bubbles in your mixture that will preferentially absorb quarks over antiquarks. Meanwhile use sphalerons to convert antiquarks into quarks outside the bubbles. Once your Higgs field has reached the desired consistency you should find that you have more quarks than antiquarks, and also that the electroweak force has separated into electromagnetic and weak forces.

Allow the resulting hot soup of quarks and gluons to continue to expand and cool for a further millionth of a second until it begins to congeal to form protons and neutrons. Allow antimatter and matter to annihilate, leaving only around one ten-billionth of the original amount of matter. Don't worry, this should be more than enough for the apple pie.

After another two minutes the mixture should have cooled below 1 billion degrees and you can begin making the first elements beyond hydrogen. Your mixture should now consist of about one neutron to every seven protons, plus a fuck-load of photons. Simmer gently at a gradually reducing temperature for about ten minutes, until nuclear fusion results in a mixture of light nuclei: about one part helium to three parts hydrogen, plus a tiny sprinkling of lithium.

Allow the hydrogen–helium mixture to continue to cool for a further 380,000 years, when, all being well, you should notice that the hot fiery mixture becomes transparent as electrons bind to hydrogen and helium nuclei to form the first neutral atoms. You can now leave the warm gases to cool unattended for a further 100 to 250 million years, so time for a nice cup of tea.

Wait over, you can start to form the first stars by collapsing large clouds of hydrogen and helium gas. In their cores, begin by converting hydrogen into helium, then helium into carbon via the triple-alpha process. You may find that these first stars are large enough to continue to fuse all the elements up to iron, which can then be spread through the mixture using supernovae.

For another 9 or so billion years, continue to make larger quantities of heavy elements in subsequent generations of stars, supernovae, and neutron star collisions, until you have a mixture with a good spread of elements from hydrogen all the way up to uranium. From this mixture, form a rocky sphere approximately 13,000 kilometres in diameter and place in the habitable zone of a yellow dwarf star. Make sure the resulting planet has sufficient quantities of hydrogen and oxygen (preferably in the form of water), carbon, and nitrogen.

Now do some biology. To be honest, I'm not at all sure about this next bit. But with a little luck after around 4.5 billion years you should end up with apples, trees, cows, and wheat, plus a

few other handy living organisms. Hopefully supermarkets will
also have spontaneously evolved by now, so go out and buy:

For the shortcrust pastry

400g plain flour, plus extra for rolling out
2 tbsp. sugar
Pinch of salt
Grated zest of a lemon
250g cold butter, cut into cubes
1 free-range egg beaten with 2 tbsp. cold water

For the filling

600g cooking apples
Juice of a lemon
50g light brown sugar plus 1 tbsp. for sprinkling
1 tsp. ground cinnamon
2 tbsp. cornflour

For finishing

1 free-range egg, beaten, for glazing
1 or 2 tbsp. demerara or light brown sugar

Now make the apple pie:

First make the pastry. Place the flour, sugar, salt, and lemon zest
into a bowl and rub in the butter until the mixture resembles
breadcrumbs. Add the beaten egg and water and stir with a
round-bladed knife until the mixture forms a dough.
Alternatively, put the dry ingredients in a food processor, briefly
whizz to combine, then add the beaten egg and water and whizz
till the pastry balls.

Wrap the pastry in cling film and chill in the fridge for 30 minutes.

Remove the pastry from the fridge, returning one-third for the lid. Roll out the remaining pastry on a lightly floured surface to 3 millimetres thick and 5 to 7 centimetres larger than the pie dish. Lift the pastry over a floured rolling pin and lower it gently into the greased pie dish.

Press the pastry firmly into the dish and up and over the sides, leaving an overhang, making sure there are no air bubbles. Chill in the fridge for 10 minutes.

Preheat the oven to 200 °C and place a baking tray in the oven to preheat.

To make the filling, peel, core, and slice the apples and place in a bowl of cold water with the juice of the lemon. Drain and pat dry.

Mix the sugar, cinnamon, and cornflour in a large bowl, then add the sliced apples and stir. Place the apple filling into the pie dish, levelling the slices but making sure that it rises above the edge of the dish. Brush the rim of the pastry with beaten egg.

Roll out the remaining ball of pastry. Cover the pie with the pastry and press the edges together firmly to seal. Using a sharp knife, trim off the excess pastry, then gently crimp all around the edge. Make a few small holes in the centre of the pie with the tip of a knife. Glaze the pie with beaten egg.

To decorate, lightly knead the pastry trimmings and roll out. Cut out some pretty shapes (leaves are traditional, but atoms and stars are also acceptable) and place on top of the pie, glazing with more egg. Chill in the fridge for 30 minutes.

Sprinkle the pie with sugar and bake in the centre of the oven for 45–55 minutes until the pastry is golden brown and the apples are tender.

Serve with cream or vanilla ice cream. Warning: filling is hot.

Acknowledgements

As I sit here in September 2020, I cannot quite believe that this book, or at least the words that will eventually go into it, is finally written. That I've got to this point is almost entirely thanks to the generosity, encouragement, patience, expertise, insight, advice, and occasional sharp shoves from dozens and dozens of people.

I am enormously grateful to the many scientists who gave their time so generously to talk to me, show me around their extra-ordinary workplaces, introduce me to their colleagues, or review parts of the manuscript, in particular: Gianpaolo Bellini, Aldo Ianni, Matthias Junker, Jennifer Johnson, Matt Kenzie, Sarah Williams, Jeffrey Hangst, Nick Manton, Joe Giaime, Karen Kinemuchi, Helen Caines, Zhangbu Xu, Lijuan Ruan, Juan Maldacena, Nima Arkani-Hamed, Joseph Conlon, Sabine Hossenfelder, Isabel Rabey, Sidney Wright, Panos Charitos, John Ellis, Sean Carroll, Günther Dissertori, and Michael Benedikt. I'd particularly like to thank David Tong and Ben Allanach for reviewing the later chapters and for gently correcting me on some of the more difficult theory bits. The book has far fewer errors thanks to them, though of course any remaining mistakes are mine. And while I've only been able to mention a few people specifically, I also owe an unquantifiable debt to my 1,400 colleagues on the LHCb experiment, as well as to the tens of thousands of people in the global scientific community and to the billions of taxpayers all over the world who fund basic, curiosity-driven research. Without them there would have been nothing to write about in the first place.

Special thanks to Graham Farmelo for his sage advice about the process of writing a book, getting me access to the hallowed halls of high theory, and for his warm encouragement. Thanks also to Neil Todd for a wonderful day as he guided me around Rutherford's old lab in Manchester.

I'm grateful to the excellent staff at the Rayleigh Library at the Cavendish Laboratory and the Science Museum Dana Research Centre and Library, particularly to the always kind and helpful Prabha Shah. Thanks also to my high-school physics teacher John Ward, both for inspiring and putting up with me as a teenager and for arranging the loan of a microscope, with the kind approval and help of Caroline Marwood.

This book would not have been possible without the support and forbearance of my boss, Val Gibson, who has been unfailingly encouraging throughout my career in physics and to whom I owe a huge debt. Thank you, Val. I'm also really grateful to my colleagues at the Science Museum, particularly Ali Boyle, from whom I learned a huge amount about how to communicate science and its history, and who gave me so many great opportunities to get better at it.

I would like to thank my brilliant agent Simon Trewin, who helped me turn what had been a very long-gestating idea into something worth writing and made this whole thing possible in the first place. Huge thanks also to Dorian Karchmar at WME in New York for doing such a fantastic job persuading US publishers to talk to a British guy about apple pie, and also to the team at WME in London, especially James Munro, Florence Dodd, and Anna Dixon.

Thank you to my editors, Ravi Mirchandani at Picador and Yaniv Soha at Doubleday. In particular, thanks to Ravi for so enthusiastically backing the arguably rather silly concept from the outset, and to Yaniv for his thoughtful and insightful feedback, which unquestionably resulted in a far better book. Thanks also to Mel Northover for turning my crappy diagrams into something far more appealing and to Amy Ryan for her forensic copyedit and for catching lots of my silly mistakes.

Finally, I would like to thank my friends and family for their love and support over the past eighteen months. To Suzie, thank you for all the mutual book-writing counselling sessions – you helped make the process feel much less lonely. I owe a special debt to my sister, Alexandra, who first suggested that perhaps I should think about writing a book almost a decade ago, a suggestion that ultimately led here. And very last but by absolutely no means least I want to thank my parents, Vicky and Robert, not only for reading and commenting on every last word of this manuscript, but for always being available when I needed to bounce ideas around, have a moan, or just needed a cup of tea and a chat. Thank you for always encouraging me to be curious; this is all your fault.

Notes

PROLOGUE

5   10,000 tonnes of liquid nitrogen: CERN, 'Cryogenics: Low temperatures, high performance', home.cern.

5   'sinister' dimension: Jon Austin, 'What is CERN doing? Bizarre clouds over Large Hadron Collider prove portals are opening', *Daily Express*, 29 June 2016, www.express.co.uk.

5   'to summon God': Sean Martin, 'Large Hadron Collider could accidentally SUMMON GOD, warn conspiracy theorists', *Daily Express*, 5 October 2018, www.express.co.uk.

7   more than €12 billion: Alex Knapp, 'How much does it cost to find a Higgs boson?', *Forbes*, 5 July 2012, www.forbes.com.

7   750-metre stretch of the accelerator: Lucio Rossi, 'Superconductivity: Its role, its success and its setbacks in the Large Hadron Collider of CERN', *Superconductor Science and Technology* 23 (2010): 034001 (17 pages).

11   'the mind of God': Stephen Hawking, *A Brief History of Time* (Bantam Books, 1988), page 175.

CHAPTER 1

14   'I do not think I shall die': Holmes, page 257.

18   'a revolution of physics and chemistry': Brock, page 104.

20   'The feeling of it': Joseph Priestley, *Experiments and Observations on Different Kinds of Air*, vol. 2 (London, 1775).

21   'to be a guinea pig': Brock, page 108.

CHAPTER 2

31   'produce the most important': Thackray, page 85.

33   'he sits at the back': Gribbin, page 7.

39   'It is to be hoped': Albert Einstein, *Investigations on the Theory of the Brownian Movement*, translated from original 1905 article by A. D. Cowper (Dover Publications, 1956), page 18.

CHAPTER 3

46   notoriously clumsy: Isobel Falconer, 'Theory and Experiment in J. J. Thomson's Work on Gaseous Discharge' (PhD dissertation, University of Bristol, 1985), page 103.

46   keep his boss from handling: Wilson, page 83.

49   'pulling their legs': Thomson, page 341.

50   'the best glassblower in England': Eve, page 34.

53   'By thunder!': Wilson, page 228.

53   'His mind was like the bow of a battleship': Chadwick, AIP interview, session 4.

55   'exceptional violence': Fernandez, page 65.

56   'quite the most incredible': Fernandez, page 73.

CHAPTER 4

65   'If, as I have reason to believe': Wilson, page 405.

65   'The hydrogen atom': Wilson, page 394.

70   'I did quite a number of quite silly': Chadwick, AIP interview, session 3.

71   'I just kept on pegging away': Chadwick, AIP interview, session 3.

74   'I don't believe it!': Hendry, page 45.

CHAPTER 5

83   the Sun blasts 383 trillion trillion watts: Sun Fact Sheet, NASA, http://nssdc.gsfc.nasa.gov/planetary/factsheet/sunfact.html.

85   'We do not argue': Kragh, page 84.

86   'I think this was the experiment': Gamow, page 15.

87   'buzzing with excitement': Gamow, page 58.

91   'were almost ready': Gamow, page 70.

92   'As soon as it grew dark': Iosif B. Khriplovich, 'The eventful life of Fritz Houtermans', *Physics Today* 45, no. 7 (1992): 29.

93   an impressive 800,000 volts: Cathcart, page 218.

95   'knew nothing about the interior': Gamow, page 136.

99   65,000 times more power: Tassoul, page 137.

99   1.2 times as heavy as the Sun: Cassé, page 82.

CHAPTER 6

113   'It's better to be interesting': Mitton, in the foreword by Paul Davies, page x.

118   he burst into Fowler's office: Mitton, page 207.

119   He later wrote: Hoyle, page 265.

119   Hoyle repeatedly descended: Hoyle, page 265.

119   with an energy of 7.19 MeV: Hoyle, page 266.

124   a beautiful colour-coded version of the periodic table: Jennifer Johnson, 'Populating the periodic table: Nucleosynthesis of the elements', *Science* 363, no. 6426 (1 February 2019): 474–78.

124   French philosopher Auguste Comte: Chown, page 56.

128   until it reaches a temperature of 100 million degrees: Frebel, page 88.

129   a sugar-cube-sized lump would weigh around a tonne: Calculated from 'a mean density of about one billion kg/m^3' in Frebel, page 92.

CHAPTER 7

144   'repugnant': Kragh, 46.

144   'the most beautiful and satisfactory explanation': Kragh, page 55.

147   Gamow's favourite Washington restaurant, Little Vienna: Alpher, AIP interview, session 1.

149   'The elements were produced in less time': Chown, page 10.

150   'about one atom every century': Kragh, page 183.

156   'The most exciting phrase': Attributed in the 'quote of the day' source code of the Fortune computer program, June 1987.

CHAPTER 8

161   'the most original and wonderful instrument': C. T. R. Wilson –
      Biographical. NobelPrize.org. Originally from *Nobel Lectures,*
      Physics 1922–1941 (Elsevier Publishing Company, 1965).

164   'Who ordered that?': Martin Bartusiak, 'Who ordered the muon?',
      New York Times, 27 September 1987.

165   'The finder of a new elementary particle': Willis Lamb, Nobel
      lecture, 12 December 1955. www.nobelprize.org.

165   'If I could remember': Robert L. Weber, *More Random Walks in*
      Science (Taylor & Francis, 1982), page 80.

168   His older brother Ben taught him to read: Gell-Mann,
      page 12.

174   'the concrete block model': Riordan, e-book location 2528.

174   'we must face the likelihood that quarks are not real': Riordan,
      e-book location 2765.

CHAPTER 9

194   'Heisenberg, how do you know': Farmelo, page 164.

CHAPTER 11

255   'That's total nonsense!': Ralph P. Hudson, 'Reversal of the parity
      conservation law in nuclear physics', in *A Century of Excellence in*
      Measurements, Standards, and Technology. NIST Special Publication
      958 (National Institute of Standards and Technology, 2001).

280   'It doesn't make a difference': Richard Feynman, 'The character of
      physical law', lecture 7, 'Seeking new laws', Messenger Lectures
      at Cornell, 1964.

CHAPTER 12

281   'With these record-shattering': CERN press release, 'LHC research
      program gets underway', 30 March 2010.

297   'ARE WE ALL GOING': Michael Hanlon, 'Are we all going
      to die next Wednesday?', *Daily Mail*, 4 September 2008,
      www.dailymail.co.uk.

297   'COLLIDER TRIGGERS': Eben Harrell, 'Collider triggers end-of-world fears', *Time*, 4 September 2008, www.time.com.

297   'The possible concern': John R. Ellis et al., 'Review of the safety of LHC collisions', *Journal of Physics G* 35, no. 11 (2008): 115004.

299   'into hospital': Pallab Ghosh, 'Popular physics theory running out of hiding places', BBC News website, 12 November 2012, www.bbc.co.uk.

299   'putting our supersymmetry theory colleagues': Pallab Ghosh, 'Popular physics theory running out of hiding places', BBC News website, 12 November 2012, www.bbc.co.uk.

300   'was actually expected': Pallab Ghosh, 'Popular physics theory running out of hiding places', BBC News website, 12 November 2012, www.bbc.co.uk.

301   'Supersymmetry is a bit late': Alok Jha, 'One year on from the Higgs boson find, has physics hit the buffers?', *Guardian*, 6 August 2013, www.theguardian.com.

CHAPTER 13

315   'We think we are beginning': Weinberg, page ix.

316   'incomparable beauty': Letter fròm Albert Einstein to Heinrich Zangger, Berlin, 26 November 1915. Translated and annotated by Bertram Schwarzschild.

317   'a lonely old fellow': Paul Halpern, *Einstein's Dice and Schrödinger's Cat* (Basic Books, 2015), page 167.

CHAPTER 14

343   around $152 billion: Alex Knapp, 'Apollo 11's 50th anniversary: The facts and figures behind the $152 billion moon landing', *Forbes*, 20 July 2019, www.forbes.com.

344   as the physicist Andrew Steele: Andrew Steele, 'Blue skies research', *Scienceogram UK*, scienceogram.org.

345   interstellar Higgs drive: Jon Butterworth, 'Impact? I want an interstellar Higgs drive please', *Guardian*, 16 July 2012, www.theguardian.com.

Bibliography

BOOKS

Ball, Philip. *Beyond Weird*. Vintage, 2018.

Brock, William H. *The Fontana History of Chemistry*. Fontana Press, 1992.

Brown, Gerald, and Chang-Hwan Lee. *Hans Bethe and His Physics*. World Scientific, 2006.

Cassé, Michael. *Stellar Alchemy: The Celestial Origin of Atoms*. Cambridge University Press, 2003.

Cathcart, Brian. *The Fly in the Cathedral*. Viking, 2004.

Chandrasekhar, S. *Eddington: The Most Distinguished Astrophysicist of His Time*. Cambridge University Press, 1983.

Chown, Marcus. *The Magic Furnace*. Jonathan Cape, 1999.

Close, Frank. *Antimatter*. Oxford University Press, 2009.

——. *The Infinity Puzzle*. Oxford University Press, 2011.

Conlon, Joseph. *Why String Theory?* CRC Press, 2016.

Crowther, J. G. *The Cavendish Laboratory 1874–1974*. Science History Publications, 1974.

Davis, E. A., and I. J. Falconer. *J. J. Thomson and the Discovery of the Electron*. Taylor & Francis, 1997.

Eve, A. S. *Rutherford: Being the Life and Letters of the Rt. Hon. Lord Rutherford, O.M.* Cambridge University Press, 1939.

Farmelo, Graham. *The Strangest Man*. Faber and Faber, 2009.

Fernandez, Bernard. *Unravelling the Mystery of the Atomic Nucleus: A Sixty Year Journey 1896–1956*. Springer, 2013.

Frebel, Anna. *Searching for the Oldest Stars*. Princeton University Press, 2015.

Gamow, George. *My World Line: An Informal Autobiography*. Viking Press, 1970.

Gell-Mann, Murray. *The Quark and the Jaguar*. Little, Brown and Company, 1994.

Green, Lucie. *15 Million Degrees: A Journey to the Center of the Sun*. Viking, 2016.

Gribbin, John. *Einstein's Masterwork*. Icon Books, 2015.

Hendry, John. *Cambridge Physics in the Thirties*. Adam Hilger, 1984.

Holmes, Richard. *The Age of Wonder*. Harper Press, 2008.

Hoyle, Fred. *Home Is Where the Wind Blows*. University Science Books, 1994.

Huang, Kerson. *Fundamental Forces of Nature: The Story of Gauge Fields*. World Scientific, 2007.

Kragh, Helge. *Cosmology and Controversy*. Princeton University Press, 1996.

Mitton, Simon. *Fred Hoyle: A Life in Science*. Aurum Press, 2005.

Pais, Abraham. *Inward Bound: Of Matter and Forces in the Physical World*. Oxford University Press, 1986.

Rickles, Dean. *A Brief History of String Theory*. Springer, 2014.

Riordan, Michael. *The Hunting of the Quark*. Simon and Schuster, 1987.

Segrè, Gino. *Ordinary Geniuses: Max Delbrück, George Gamow, and the Origins of Genomics and Big Bang Cosmology*. Viking, 2011.

Tassoul, Jean-Louis, and Monique Tassoul. *A Concise History of Solar and Stellar Physics*. Princeton University Press, 2004.

Thackray, Arnold. *John Dalton: Critical Assessments of His Life and Science*. Harvard Monographs in the History of Science. Harvard University Press, 1972.

Thomson, J. J. *Recollections and Reflections*. G. Bell and Sons, 1936.

Vilbert, Douglas A. *The Life of Arthur Stanley Eddington*. Thomas Nelson and Sons, 1956.

Weinberg, Steven. *Dreams of a Final Theory*. Vintage, 1993.

Wilson, David. *Rutherford, Simple Genius*. Hodder and Stoughton, 1983.

OTHER

BBC Radio 4, *In Our Time: John Dalton*, 26 October 2016.

Interview of James Chadwick by Charles Weiner on 20 April 1969, Niels Bohr Library & Archives, American Institute of Physics. www.aip.org.

Interview of Ralph Alpher by Martin Harwit on 11 August 1983, Niels Bohr Library & Archives, American Institute of Physics. www.aip.org.

Interview of Carl Anderson by Charles Weiner on 30 June 1966, Niels Bohr Library & Archives, American Institute of Physics. www.aip.org.

Index

(Page references in *italics* refer to illustrations.)